# Sulfate Attack on Concrete

# Modern Concrete Technology Series

A series of books presenting the state-of-the-art in concrete technology

## Series Editors

Arnon Bentur
*National Building Research Institute*
*Technion–Israel Institute of Technology*
*Technion City*
*Haifa 32 000*
*Israel*

Sydney Mindess
*Office of the President*
*University of British Colombia*
*6328 Memorial Road*
*Vancouver, B.C.*
*Canada V6T 1Z2*

# Sulfate Attack on Concrete

**Jan Skalny, Jacques Marchand
and Ivan Odler**

**CRC Press**
Taylor & Francis Group
Boca Raton London New York

CRC Press is an imprint of the
Taylor & Francis Group, an **informa** business

A SPON PRESS BOOK

CRC Press
Taylor & Francis Group
6000 Broken Sound Parkway NW, Suite 300
Boca Raton, FL 33487-2742

First issued in paperback 2019

© 2002 Jan Skalny, Jacques Marchand and Ivan Odler
CRC Press is an imprint of Taylor & Francis Group, an Informa business

No claim to original U.S. Government works

ISBN-13: 978-0-419-24550-6 (hbk)
ISBN-13: 978-0-367-41714-4 (pbk)

*British Library Cataloguing in Publication Data*
A catalogue record for this book is available from the British Library

*Library of Congress Cataloging in Publication Data*
A catalogue record has been requested

**Visit the Taylor & Francis Web site at
http://www.taylorandfrancis.com**

**and the CRC Press Web site at
http://www.crcpress.com**

# Contents

**4 Sulfate attack**    43

# Preface

We consider it to be a great honor to be chosen by Professors A. Bentur and S. Mindess to prepare this book for publication by Spon Press and it was not without trepidation that we accepted the task to summarize the available knowledge on the effect of sulfates on concrete. We were hesitant because the mechanistic issues of sulfate attack on concrete are complex and sometimes controversial.

The multidimensionality of the sulfate attack issues becomes obvious when one realizes the variability of the environmental conditions under which concrete is used, of the complex chemistry and mineralogy of the concrete components, and of the less-than-well-defined processing conditions used in making concrete to be used in a variety of structures. In discussing sulfate attack mechanisms, one has to deal, among others, with issues of chemical and mineralogical composition of the aggressive species, properties of amorphous to crystalline reaction products, variability and limited controllability of the reaction (curing) conditions, and to consider difficult issues related to best testing methodology, standardization, prediction of service life by modeling, etc. All the above make consolidation of the existing knowledge a considerable challenge.

The book is prepared for an audience consisting of students of materials engineering, construction practitioners, and researchers. As the loss of durability of concrete almost always involves chemical and physico-chemical processes, it may be difficult for some readers to follow the intricate details of sulfate attack mechanisms. To overcome this, we included a few chapters summarizing the basics of cement hydration and concrete deterioration. It is entirely possible that for some, the book contains excessive information; others may consider the given details to be inadequate. There will be voices disagreeing with our interpretation of the "average" literature data. To these we can only say that we did our best to consolidate the available data in an understandable manner and we suggest that, if needed, the readers visit the references given at the end of each chapter.

We hope that this book will be considered a good introduction to the scientific and practical intricacies of sulfate attack mechanisms, as well as becoming an impetus to younger researchers to develop a *romance* with engineering materials.

# Acknowledgments

The authors would like to thank the numerous colleagues who helped in realizing this complex project. First of all, we would like to acknowledge the invitation by Professors A. Bentur and S. Mindess, the Editors of this Spon Press series of books on concrete technology, to prepare this manuscript for publication. Although an honor, the consolidation of all the available controversial data was not without difficulties and frustrations.

We are most thankful to Messrs P.W. Brown, S. Diamond, J. Gebauer, V. Johansen, B. Mather, K.L. Scrivener, H.F.W. Taylor, N. Thaulow, M. Thomas and unnamed others for most useful and critical discussions on mechanistic aspects of various forms of sulfate attack. We would like to acknowledge our gratefulness to J.J. Beaudoin, N.J. Crammond, S. Diamond, G. Frohnsdorff and E.M. Gartner for their critical review of parts of the manuscript, and to N.J. Crammond and coworkers for contributing the case study on thaumasite. Special *in memoriam* thanks are due to C.D. Lawrence who made us available his unpublished bibliography of "DEF" literature and J. Jambor who left with us his unpublished report on sulfate attack.

We would also like to acknowledge the authors of the high-quality photographic material used throughout the book (obtained from Messrs M. Alexander, S. Badger, C. Fourie, U. Hjorth Jakobsen, S. Sahu, P. Stutzman, N. Thaulow and M. Thomas), as well as photographic and other material obtained from CEMBUREAU (Belgium), G.M. Idorn Consult (Denmark), BRE – Building Research (United Kingdom), R.J. Lee Group (USA), The Erlin Company (USA), Portland Cement Association (USA), and other sources. Our thanks are also due to J. Parent (Laval University, Quebec, Canada) for preparation of the graphic material.

Our deep thanks are due to the staff of Spon Press, specifically Marie-Louise Logan and Richard Whitby, for their dedicated work and professionalism. Finally, important thanks are also due to our spouses – Magdalena,

Marie-Helene and Marika – for exercising patience that some of us cannot reciprocate with.

Jan Skalny (jpskalny@aol.com)
Jacques Marchand (jacques.marchand@gci.ulaval.ca)
Ivan Odler (ivanodler@aol.com)
January 2001

# 1   Introduction

*Portland clinker-based concrete* is the most important and versatile construction material used. It is an extremely complex *composite material* and, considering its chemical, physical, and microstructural intricacy, it is also a very forgiving material: in spite of severe abuse by Man and Nature, most of the immense amounts of Portland clinker-based concrete used World over are performing its intended functions surprisingly well.

However, this generally good performance of concrete is not a satisfactory excuse for improper or inadequate utilization by Man of the available knowledge generated during the past 100 or more years. To the contrary, the cost of repair of deteriorated concrete and its possible replacement, not speaking about the societal cost of expensive litigations and other unnecessary expenses, more than justifies investment into better understanding of the nature of concrete and its performance in the environment it is used. This book is meant to be a humble contribution to dissemination of available information about basic aspects of concrete material science and, more specifically, about proper treatment of both fresh and hardened concrete to assure long-lasting durability of concrete structures in sulfate-bearing environment.

Man abuses concrete by:

- use of wrong or marginal concrete materials and improper mix proportions;
- inappropriate use of concrete mix compositions in structures exposed to harsh environment and structural design unsuitable for the given environmental exposure;
- curing or heat treatment in conflict with chemistry and physics of concrete microstructure development;
- wrong placement and finishing procedures; and
- lack of maintenance.

Nature causes additional challenges. Of these, the most important examples are:

- environmental conditions (extreme temperatures, temperature and humidity fluctuations);
- access to concrete of chemical species capable of reacting with concrete components (atmospheric pollution, ground water components, industrial waste, chlorides from sea water or de-icing salts); and
- instability of many siliceous aggregates in the alkaline environment of Portland cement concrete (e.g. rock components containing amorphous silica); dolomitic limestone.

To produce concrete of highest quality and better than expected service life, both the challenges of Nature and the inadequacies of Man have to be taken into consideration. This can be done by:

- improved utilization of the basic chemical and physical principles governing the formation and destruction of cement-based materials;
- designing concrete mixes and structures for the specific environment of use; and
- proper production, placement, and maintenance.

All these tasks require quality education of those involved, including the management, research and engineering, and the actual construction staffs.

Cement production and consumption are considered to be important indicators of economic growth. To give the reader an appreciation of the size of the cement business world-wide, an overview of the US and World consumption and the top ten World producers are given in Tables 1.1 and 1.2 (PCA 2000; CEMBUREAU 2000). Consumption of concrete obviously follows cement consumption; considering these large amounts, concrete is clearly the most used construction material.

*Sulfate attack* is a generic name for a set of complex and overlapping chemical and physical processes caused by reactions of numerous cement components with sulfates originating from external or internal sources. For the purpose of the following discussion, the term "cement components" will refer to both the actual clinker minerals, such as calcium silicates and calcium

*Table 1.1* US and World cement consumption of Portland clinker-based hydraulic cements (in millions of metric tons).

| Year | 1976 | 1980 | 1984 | 1988 | 1992 | 1994 | 1995 | 1996 | 1997 | 1998 | 1999 |
|------|------|------|------|------|------|------|------|------|------|------|------|
| USA | 64 | 67 | 72 | 81 | 73 | 82 | 83 | 87 | 93 | 99 | 105 |
| World | 754 | 878 | 934 | 1,116 | 1,238 | 1,366 | 1,438 | 1,444 | 1,482 | 1,537 | 1,603 |

*Table 1.2* Top ten world producers of hydraulic cements (in millions of metric tons). Includes exported clinker.

| Country | 1988 | 1990 | 1992 | 1994 | 1996 | 1997 | 1998 | 1999[e] |
|---------|------|------|------|------|------|------|------|---------|
| China | 210 | 211 | 308 | 421 | 491 | 512 | 536 | 573 |
| Japan | 79 | 87 | 96 | 97 | 98 | 96 | 83 | 83 |
| USA | 70 | 70 | 70 | 78 | 79 | 83 | 84 | 85[e] |
| Russia[e] | 84 | 83 | 62 | 37 | 28 | 27 | 26 | 26 |
| India | 45 | 54 | 54 | 58 | 75 | 85 | 87 | 99 |
| Germany | 37 | 35 | 33 | 36 | 31 | 31 | 36 | 37 |
| Italy | 39 | 41 | 41 | 33 | 34 | 35 | 36 | 36 |
| Korea | 29 | 34 | 43 | 55 | 59 | 60 | 48 | 49 |
| Brazil | 25 | 26 | 24 | 25 | 35 | 38 | 40 | 40 |
| Thailand | 12 | 18 | 22 | 33 | 40 | 42 | 30 | 35 |

Total world production: 1,603 millions of metric tons (1999)
e = estimate

aluminates, and supplementary materials present in modern cements, such as slag, fly ash, calcined kaolin, and microsilica.

There is some confusion in the literature and the technical community regarding the definition of sulfate attack. For some, the term means only the process of possible expansion caused by formation of ettringite from external source of sulfate with the $C_3A$ present in the used cement. Others do not consider damage caused by formation and recrystallization of thenardite to/from mirabelite to be sulfate attack, and call it physical attack or salt crystallization. The literature, including some standards, gives several "variations on the theme." In our opinion, and this is supported by credible scientific data, *sulfate attack* is a complex set of processes that cannot be easily divided into physical versus chemical or calcium- versus magnesium- versus sodium-sulfate attack. Depending on the nature of the concrete components, the concrete processing conditions, the local macro- and micro-environments, and the form, concentration, and nature of the sulfates in contact with the concrete, more than one of these complicated chemical and physical phenomena may occur simultaneously. This complexity is well known for years and has been discussed, among others, by Lerch (1945), Thorvaldson (1952), Eitel (1957), Kalousek *et al.* (1976), Mehta (1992, 2000), St John *et al.* (1998), Taylor (1997), Skalny and Pierce (1999), Hime and Mather (1999), Mather (2000), and many others.

The sulfate anion that reacts with cement components of concrete to cause damage is originally present in the deteriorating system mostly in the form of highly-soluble alkali ($Na_2SO_4$, $K_2SO_4$) or alkali earth ($CaSO_4 \cdot 2H_2O$, $MgSO_4$) salts or, less frequently, originates from the oxidation of pyrite in the aggregate, from fertilizers, or from various forms of industrial waste. Since cement and concrete are chemically and microstructurally highly complex composites, and the ionized sulfates are often associated with more than one or even several different cations, the chemical processes that lead to eventual deterioration of concrete properties are highly complex, interdependent, and overlapping. In addition to the above chemical reasons, these reactions depend on the environmental exposure of the particular concrete structure, including access of moisture, rate of water evaporation, and temperature changes.

A few examples of damage caused by various sulfate attack mechanisms are presented in Figures 1.1–1.4.

Deterioration of concrete by sulfates has been historically assessed in numerous ways, neither of which gives adequate – meaning reproducible and accurate – results under all conditions. Such assessment techniques include visual evaluation, wear rating, loss of mass, hardness, compressive or tensile strength, dynamic modulus of elasticity, and volume instability, and are usually recommended by codes and standards (e.g. ASTM 1995a, 1995b; Hobbs 1998). As the mechanisms of concrete deterioration due to sulfates are multi-faceted, it is now clear that sulfate attack cannot be fully characterized by a single indirect test (Clifton *et al.* 1999; Hooton 1999; Skalny and Pierce 1999; Taylor 2000). Presently used tests are deemed to be indirect because they do not take into consideration the actual cause of deterioration but only measure the physical or mechanical consequence of the damage. For a partial list of standards and test methods pertaining to sulfate attack in concrete see Table 1.3.

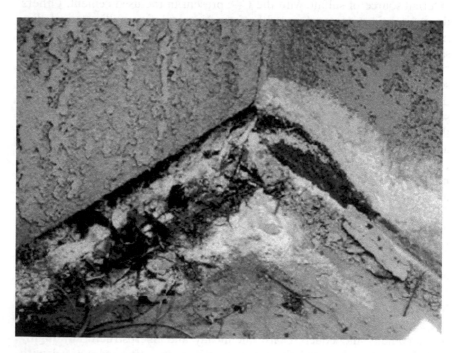

*Figure 1.1* Deposition of sulfate-bearing efflorescing material at the exposed concrete foundation of a residential home (Photo: J. Skalny).

*Table 1.3* List of selected standards and test methods pertaining to sulfate attack in concrete and related material.

ASTM Designation E 150 – *Specification for Portland cement*

ASTM Designation C 452 – *Test method for potential expansion of Portland cement mortars exposed to sulfate*

ASTM Designation C 632 – *Standard practice for developing accelerated tests to aid prediction of the service life of building components and materials*

ASTM Designation C 1012 – *Test method for length change of hydraulic-cement mortars exposed to a sulfate solution*

ASTM Designation C 1157M – *Performance specification for blended hydraulic cement*

ACI 201 (1998) "Guide to Durable Concrete", *ACI Manual of Concrete Practice: Part 1*, ACI Farmington Hill, MI

Uniform Building Code (1997) *Concrete*, vol. 2, Chapter 19.

British Standard Institution, BS 5328 (1997) "Guide to specifying concrete", *Concrete – Part 1*.

British Standards Institution (1997) "Cement – Part 1: Composition, specifications and conformity criteria of common cements", Pr ENV 197-1. Document 97/103566, Committee B/516.

British Standards Institution (1997) "Sulfate-resisting cements", Pr ENV 197-X. BSI Document 97/103303, Committee B/516/6.

European Standard (draft) (1998) *Common Rules for Precast Concrete Products*, CEN TC 229, April.

German Committee for Reinforced Concrete (1989) *Recommendation on the Heat Treatment of Concrete* (in German), Berlin, September.

BRE Digest 363 (1996) *Sulfate and acid attack on concrete in the ground*, British Research Establishment, Garston, Watford, UK.

Hobbs, D.W. (1998) *Minimum Requirements for Durable Concrete: Carbonation- and Chloride-induced Corrosion, Freeze-thaw Attack and Chemical Attack*, British Cement Association.

Spooner, D.C. (1995) "The selection of Portland cements to British standards and on European prestandard ENV-197-1", *The Structural Engineer* **73**(20): 17–19.

*Figure 1.2* Visible surface deterioration of concrete curbs exposed to Na- and Mg-
     sulfates present in ground water. Efflorescing material identified as
     sodium sulfate (Photo: J. Skalny).

It should be also noted that some of the tests used for assessment of
concrete durability are inadequate measures of the remaining service life.
Compressive strength is a typical example; its inadequacy in characterizing
the degree of concrete deterioration at any given time was recognized long
time ago and was recently discussed (Mehta 1997; Neville 1998; Jambor 1998).

*Figure 1.3* Damaged and undamaged railroad ties (Photo courtesy of N. Thaulow).

*Figure 1.4* Laboratory concrete samples attacked by sulfuric acid; paste portion readily soluble: (a) Sample with dolomitic (acid-soluble) aggregate; and (b) sample with silicious (insoluble) aggregate (Photographs courtesy of C. Fourie and M. Alexander).

*Table 1.4* Requirements for concrete exposed to sulfate-containing solutions.

| Sulfate exposure | Water-soluble sulfate (SO₄) in soil, (% by weight) | Sulfate (SO₄) in water (in ppm) | Cement type | Maximum w/cm, by weight (for normal-weight aggregate concrete)[1] |
|---|---|---|---|---|
| Negligible | 0.00–0.10 | 0–150 | – | – |
| Moderate[2] | 0.10–0.20 | 150–1,500 | II, IP (MS), IS (MS) | 0.50 |
| Severe | 0.20–2.00 | 1,500–10,000 | V | 0.45 |
| Very severe | >2.00 | >10,000 | V plus pozzolan[3] | 0.45 |

*Source:* "Guide to Durable Concrete" (ACI 201-2R-92). Reprinted with permission by the American Concrete Society

*Notes*:
1  A lower w/cm may be required for low permeability or protection against corrosion or freezing and thawing
2  Includes sea water
3  Pozzolan that has been determined by test or service record to improve sulfate resistance when used in concrete containing Type V cement

*Table 1.5* Proposed requirements to protect against damage to concrete by sulfate attack by external sources of sulfate (ACI Committee 201).

| Severity of potential exposure | Water-soluble sulfate (SO₄) in soil (in % by mass) | Sulfate (SO₄) in water (in ppm) | Maximum water-to-cementitious material ratio (by mass) | Cementitious materials requirements |
|---|---|---|---|---|
| Class 0 Exposure | 0.00 to 0.10 | 0 to 150 | No special requirement for sulfate resistance | No special requirement for sulfate resistance |
| Class 1 Exposure | More than 0.10 to less than 0.20 | More than 150 to less than 1,500 | 0.50 | C 150 Type II or eqivalent[#] |
| Class 2 Exposure | 0.20 to less than 2.0 | 1,500 to less than 10,000 | 0.45 | C 150 Type V or equivalent[#] |
| Class 3 Exposure | 2.0 or greater | 10,000 or greater | 0.40 | C 150 Type V plus pozzolan or slag[#] |
| Sea water Exposure | | | 0.45 | C 150 Type II with maximum 10% C3A or equivalent[#] |

[#] For detailed explanation see ACI 201 – *A Guide to Durable Concrete*

Sulfate attack on concrete is known for many decades, and it's scientific and engineering consequences have been studied by many well-established institutions (PCA, NBS, Bureau of Reclamation, Cement and Concrete Association) and individuals (see publications e.g. HRB 1966; Swenson 1968; ACI 1982; Marchand and Skalny 1999; Erlin 1999). However, changing cement and concrete processing conditions, changed properties of modern cements, as well as the availability of new experimental and computational techniques, all call for re-evaluation of the existing knowledge on the mechanistic aspects of these reactions and of preventive measures. The present-day ACI and UBC requirements for concrete exposed to sulfate containing solutions are summarized in Table 1.4 (UBC 1997). Changes that are presently considered by ACI Committee 201 – Concrete Durability are given in Table 1.5. The primary proposed change is introduction of 0.4 w/cm for most severe sulfate exposure.

## REFERENCES

ACI (1982) George Verbeck Symposium on *Sulfate Resistance of Concrete*, American Concrete Institute, SP-77.

ASTM (1995a) ASTM Designation C 1012, "Standard test method for length change of hydraulic cement mortar exposed to sulfate solutions", ASTM, Philadelphia.

ASTM (1995b) ASTM Designation C 452, "Standard test method for potential expansion of Portland-cement mortars exposed to sulfate", ASTM, Philadelphia.

CEMBUREAU (2000) Cembureau EL/AD Aug-2000.

Clifton, J.R., Frohnsdorff, G. and Ferraris, C. (1999) "Standards for evaluating the susceptibility of cement-based materials to external sulfate attack", in J. Marchand and J. Skalny (eds) *Materials Science of Concrete Special Issue: Sulfate Attack Mechanisms*, The American Ceramic Society, Westerville, OH, pp. 337–356.

Erlin, B. (ed.) (1999) *Ettringite – The Sometimes Host of Destruction*, American Concrete Institute, SP-177, 265 pp.

Eitel, W. (1957) "Recent investigations of the system lime-alumina-calcium sulfate-water and its importance in building research problems", *Journal of the American Concrete Institute* **28**(7): 679–697.

Hime, W.G. and Mather, B. (1999) "'Sulfate attack,' or is it?", *Cem. Concr. Res.* **29**: 789–791.

Hobbs, D.W. (1998) *Minimum Requirements for Durable Concrete*, British Cement Association, United Kingdom.

Hooton, R.D. (1999) "Are sulfate resistance standards adequate?", in J. Marchand and J. Skalny (eds) *Materials Science of Concrete Special Issue: Sulfate Attack Mechanisms*, The American Ceramic Society, Westerville, OH, pp. 357–366.

HRB (1966) Symposium on *Effects of Aggressive Fluids on Concrete*, Highway Research Record 113, HRB, Washington, D.C.

Jambor, J. (1998) "Sulfate corrosion of concrete", unpublished manuscript summarizing his views on sulfate durability of concrete. (Dr Jambor passed away in May 1998.)

Kalousek, G.L., Porter, L.C. and Harboe, E.M. (1976) "Past, present, and potential developments of sulfate-resisting concretes", *J. of Testing and Evaluation* **4**(5) (September): 347–354.

Lerch, W. (1945) "Effect of SO$_3$ content of cement on durability of concrete", PCA Pamphlet #0285.

Marchand, J. and Skalny, J. (eds) (1999) *Materials Science of Concrete Special Volume: Sulfate Attack Mechanisms*, The American Ceramic Society, Westerville, OH, 371pp.

Mather, B. (2000) "Sulfate attack on hydraulic-cement concrete", presented at ACI/CANMET mtg. in Barcelona, Spain, June.

Mehta, P.K. (1992) "Sulfate attack on concrete – a critical review", in J. Skalny (ed.) *Materials Science of Concrete*, vol. III, The American Ceramic Society, Westerville, OH, pp. 105–130.

Mehta, P.K. (1997) "Durability – critical issues for the future", *Concrete International* **19**(7): 27–33.

Mehta, P.K. (2000) "Sulfate attack on concrete: separating the myth from reality", *Concrete International* **22**(8): 57–61.

Neville, A. (1998) "A 'new' look at high-Alumina cement," *Concrete International* **20**(8): 51.

PCA (2000) US Cement Industry Fact Sheet, 16th edn, PCA Economic Research.

Skalny, J. and Pierce, J. (1999) "Sulfate attack issues", in J. Marchand and J. Skalny (eds) *Materials Science of Concrete Special Issue: Sulfate Attack Mechanisms*, The American Ceramic Society, Westerville, OH, pp. 49–63.

St John, D.A., Poole, A.B. and Simms, I. (1998) *Concrete Petrography*, Arnold, London.

Swenson, E.G. (ed.) (1968) *Performance of Concrete: Resistance of Concrete to Sulfate and Other Environments*, University of Toronto Press.

Taylor, H.F.W. (1997) *Cement Chemistry*, 2nd edn, Thomas Telford Publishing, London.

Taylor, H.F.W. (2000) Presentation at the annual meeting of the American Ceramic Society, Cincinnati, OH, May.

Thorvaldson, T. (1952) "Chemical aspects of the durability of cement products", in *Proceedings of the 3rd Int. Symposium on the Chemistry of Cement*, CCA, London, pp. 436–466.

Uniform Building Code (1997) "Concrete", Chapter 19, in *Structural Engineering Design Provisions*, vol. 2, pp. 2-97–2-183.

# 2 Chemistry and physics of cement paste

## 2.1 CONCRETE COMPONENTS

Concrete is an inorganic composite material formed, in its simplest form, from a simple reactive binder, an inert filler, and water. In reality, modern concrete is a complex material typically made of a form of hydraulic cement, fine and course aggregate, mineral and chemical admixtures, and mix water. The structural properties of plain concrete depend primarily on the chemical reactions between the cement, water and other mix constituents, as well as on the spatial distribution and homogeneity of the concrete components. The chemistry, structure, and mechanical performance of the products of the hydration reactions in concrete are, in turn, influenced by the production processes and the environmental conditions prevailing during the production of concrete. Thus, in designing concrete for service in a specific environment, not only the concrete materials *per se*, but also the processing techniques and environments of use have to be taken into account. This fact is sometimes neglected in engineering practice.

### 2.1.1 Hydraulic cements

Modern *hydraulic* cements, cements capable of developing and maintaining their properties in moist environment, are based either on calcium aluminates (calcium aluminate or high-alumina cements) or on calcium silicates (Portland-clinker based cements). In this work, focus will be entirely on Portland cements and their modifications.

Portland cements and other Portland clinker-based hydraulic cements are produced by inter-grinding Portland cement clinker with limited amount of calcium sulfate (gypsum, hemihydrate, anhydrite; industrial by-products) and, often, with one or several mineral components such as granulated blast furnace slag, natural or artificial pozzolan, and/or limestone. Cement clinker is a precursor produced by heat treatment of a raw meal typically containing sources of lime, silica, alumina and ferrite. The main reactive components of

*Table 2.1* Clinker components: chemical and mineralogical names, oversimplified chemical formulas[#], and abbreviations*.

| Compound | Chemical formula | Abbreviations |
|---|---|---|
| Alite, tricalcium silicate | $Ca_3SiO_5$ | $C_3S$ |
| Belite, dicalcium silicate | $Ca_2SiO_4$ | $\beta\text{-}C_2S$ |
| Tricalcium aluminate | $Ca_3Al_2O_6$ | $C_3A$ |
| Tetracalcium alumino-ferrite or ferrite solid solution | $Ca_2(Al_xFe_{1-x})_2O_5$ | $C_4AF$, Fss |
| Free lime | CaO | C |
| Periclase, free magnesia | MgO | M |
| Arcanite | $K_2SO_4$ | $K\bar{S}$ |
| Thenardite | $Na_2SO_4$ | $N\bar{S}$ |
| Aphthitalite | $K_3Na(SO_4)_4$ | $K_3N\bar{S}_4$ |
| Calcium langbeinite | $K_2Ca_2(SO_4)_3$ | $KC_2\bar{S}_3$ |
| Gypsum | $CaSO_4 \cdot 2H_2O$ | $C\bar{S}H_2$ |
| Hemihydrate | $CaSO_4 \cdot 0.5H_2O$ | $C\bar{S}H_{0.5}$ |
| Anhydrite | $CaSO_4$ | $C\bar{S}$ |

[#] For more accurate and detailed information, see Taylor (1997)
*Cement chemical abbreviations: C – CaO, S – $SiO_2$, A – $Al_2O_3$, F – $F_2O_3$, M – MgO, K – $K_2O$, N – $Na_2O$, $\bar{S}$ – $SO_3$, C – $CO_2$

cement clinker are calcium silicates, aluminates and ferrites, plus minor components such as free oxides lime and periclase, and various alkali sulfates. Table 2.1 summarizes some primary clinker components and their chemical abbreviations. Note that the actual chemical compositions of many of the listed compounds are much more complex (Taylor 1997).

Reaction of individual clinker minerals and other cement components with mix water proceeds under given environmental conditions as a complex set of interdependent reactions. It is not only the chemical composition of the anhydrous compounds present, but also their "reactivity" and the composition of the liquid phase (pore solution) at any given moment, that control the direction and kinetics of the concrete setting and hardening. This "reactivity" depends, among other factors, on the crystal structure of the individual compounds (concentration and form of crystal defects) and on the temperature of hydration. Presence of chemical admixtures and reactivity of "inert" aggregate play an additional role. Typical compositions of Portland cement, fly ash, slag, and microsilica are given in Table 2.2.

## 2.1.2 Aggregates

Aggregate is the most voluminous component of concrete. Depending on the desired concrete properties, primarily strength but also durability and other properties, the mass of aggregate in concrete represents about 3.5 (for

*Table 2.2* Typical compositions of cement clinker and cement components (mass per cent).

| Oxide | Abbreviation | Cement clinker | Fly ash | GBFS | Microsilica |
|---|---|---|---|---|---|
| CaO | C | 64–65 | 1–20 | 30–50 | |
| $SiO_2$ | S | 20–22 | 10–50 | 25–45 | 90–98 |
| $Al_2O_3$ | A | 4–7 | 10–30 | 5–13 | trace |
| $Fe_2O_3$ | F | 3–5 | 1–15 | <1 | trace |
| MgO | M | 1–4 | 1–4 | 1–20 | |
| $SO_3$ | $\bar{S}$ | 0.3–1.5 | 0–5 | <3 | |
| $Na_2O$ | N | 0.1–1.5 | 0–4 | <2 | trace |
| $K_2O$ | K | 0.1–1.5 | 0–3 | <2 | trace |

high-strength) to 7.5 (for low-strength) times the amount of cement used to bind it into a solid concrete composite. This large proportion of aggregate used in concrete calls for the aggregate to possess characteristics that will give both the fresh and hardened concrete the desired engineering properties. Fine and course aggregates, whether natural or artificial, have to be selected to enable adequate workability, compaction and finishability of fresh/plastic concrete, as well as strength, elastic modulus and volume stability, among others, of hardened concrete.

The quality of any aggregate, in addition to its chemical and mineralogical nature, depends on its prior exposure to the environment and during processing. All above factors determine the microstructure of the aggregate at the time of use. An illustration of the interdependence of the aggregate properties and its microstructure is schematically given in Figure 2.1.

Microstructure of aggregate is of particular interest from the point of view of concrete durability. Surface quality, density, porosity, permeability, and chemical reactivity of an aggregate with paste and pore solution are of particular importance in chemical attack, and are of increasing importance with increasing permeability of the concrete. More often than not, the used aggregate has limited effect on chemical durability of concrete; it is usually the paste quality that controls the chemical resistance of concrete. However, there are cases where aggregate quality may affect the chemical processes of deterioration, an example being the alleged effect of aggregate composition on DEF-type of internal sulfate attack (e.g. Lawrence 1995).

Although related to total porosity, the strength of concrete is not, in itself, an adequate measure of durability. Thus, use of "strong" aggregate instead of *quality* aggregate is not recommended; durable concrete requires not only quality but also an intelligent use of the particular aggregate in a way specific to the structure's design in the given environment – a systems approach.

For more detailed information about aggregate types and their quality, the reader is advised to check specialized literature (e.g. Mehta and Monteiro 1993; Alexander 1998).

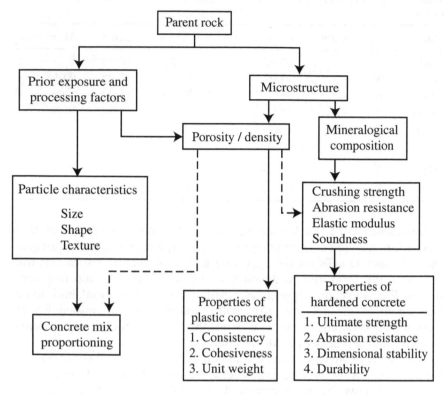

*Figure 2.1* Interdependence of aggregate microstructure and properties.
*Source*: *Concrete*, 2nd edn, Mehta–Monteiro, McGraw Hill, 1993

### 2.1.3 Mineral and chemical admixtures

Chemical and mineral admixtures are accepted components of modern concrete. They are used to enable easier processing of fresh concrete, to better the properties of hardened concrete in a structure, and to improve concrete durability and extend its service life. If used properly, admixtures can improve the economy of concrete making and enable use of concrete in new applications. Tables 2.3, 2.4 (Mehta and Monteiro 1993) and 2.5 summarize the most important properties of common admixtures.

Since admixtures affect the microstructure of the hardened concrete matrix, they may dramatically influence concrete durability. This is done primarily through their effect on overall paste porosity and permeability to water containing dissolved chemical species. Although admixtures are typically used to decrease porosity and permeability, if misused, admixtures – whether mineral or chemical – can lead to unwanted problems. Their proper use is most important also in structures potentially exposed to external sulfates.

*Table 2.3* Commonly used chemical admixtures.

| Primary function | Principal active ingredients/ASTM specification | Side effects |
|---|---|---|
| *Water-reducing* | | |
| Normal | Salts, modifications and derivatives of lignosulfonic acid, hydroxylated carboxylic acids, and polyhydroxy compounds. ASTM C 494 (Type A) | Lignosulfonates may cause air entrainment and strength loss; Type A admixtures tend to be set retarding when used in high dosage |
| High range | Sulfonated naphthalene or melamine formaldehyde condensates. ASTM C 494 (Type F) | Early slump loss; difficulty in controlling void spacing when air entrainment is also required |
| *Set-controlling* | | |
| Accelerating | Calcium chloride, calcium formate, and triethanolamine. ASTM C 494 (Type C) | Accelerators containing chloride increase the risk of corrosion of the embedded metals |
| Retarding | Same as in ASTM Type A; compounds such as phosphates may be present. ASTM C 494 (Type B) | |
| *Water-reducing and set-controlling* | | |
| Water-reducing and retarding | Same as used for normal water reduction. ASTM C 494 (Type D) | See Type A above |
| Water-reducing and accelerating | Mixtures of Types A and C. ASTM C 494 (Type E) | See Type C above |
| High-range water-reducing and retarding | Same as used for Type F with lignosulfonates added. ASTM C 494 (Type G) | See Type F above |
| *Workability-improving* | | |
| Increasing consistency | Water-reducing agents [e.g. ASTM C 494 (Type A)] | See Type A above |
| Reducing segregation | (a) Finely divided minerals (e.g. ASTM C 618) | Loss of early strength when used as cement replacement |
| | (b) Air-entrainment surfactants (ASTM C 260) | Loss of strength |

*Table 2.3  (continued)*

| Primary function | Principal active ingredients/ ASTM specification | Side effects |
|---|---|---|
| *Strength-increasing* | | |
| By water-reducing admixtures | Same as listed under ASTM C 494 (Types A, D, F, and G) | See Types A and F above |
| By Pozzolanic and cementitious admixtures | Same as listed under ASTM C 618 and C 989 | Workability and durability may be improved |
| *Durability-improving* | | |
| Frost action | Wood resins, protein-aceous materials, and synthetic detergents (ASTM C 260) | Strength loss |
| Thermal cracking Alkali-aggregate expansion Acidic solutions Sulfate solutions | Fly ashes, and raw or calcined natural pozzolans (ASTM C 618); granulated and ground iron blast-furnace slag (ASTM C 989); condensed silica fume; rice husk ash produced by controlled combustion. (High-calcium and high-alumina fly ashes, and slag-Portland cement mixtures containing less than 60% slag may not be sulfate resistant.) | Loss of strength at early ages, except when highly pozzolanic admixtures are used in conjunction with water-reducing agents |

*Source*:  *Concrete*, 2nd edn, Mehta–Monteiro, McGraw Hill 1993, pp. 286–287, Table 8.7

*Table 2.4*  Commonly used mineral admixtures.

| Classification | Chemical and mineralogical composition | Particle characteristics |
|---|---|---|
| *Cementitious and pozzolanic* | | |
| Granulated blast-furnace slag (cementitious) | Mostly silicate glass containing mainly calcium, magnesium, aluminum, and silica. Crystalline compounds of melilite group may be present in small quantity | Unprocessed material is of sand size and contains 10–15% moisture. Before use it is dried and ground to particles less than 45 µm (usually about 500 m$^2$/kg Blaine). Particles have rough texture |

| | | |
|---|---|---|
| High-calcium fly ash (cementitious and pozzolanic) | Mostly silicate glass containing mainly calcium, magnesium, aluminum, and alkalies. The small quantity of crystalline matter present generally consists of quartz and $C_3A$; free lime and periclase may be present; $C\bar{S}$ and $C_4A_3\bar{S}$ may be present in the case of high-sulfur coals. Unburnt carbon is usually less than 2% | Powder corresponding to 10–15% particles larger than 45 µm (usually 300–400 $m^2$/kg Blaine). Most particles are solid spheres less than 20 µm in diameter. Particle surface is generally smooth but not as clean as in low-calcium fly ashes |
| *Highly active pozzolans* | | |
| Condensed silica fume | Consists essentially of pure silica in noncrystalline form | Extremely fine powder consisting of solid spheres of 0.1 µm average diameter (about 20 $m^2$/g surface area by nitrogen adsorption) |
| Rice husk ash | Consists essentially of pure silica in noncrystalline form | Particles are generally less than 45 µm but they are highly cellular (about 60 $m^2$/g surface area by nitrogen adsorption) |
| *Normal pozzolans* | | |
| Low-calcium fly ash | Mostly silicate glass containing aluminum, iron, and alkalies. The small quantity of crystalline matter present generally consists of quartz, mullite, sillimanite, hematite, and magnetite | Powder corresponding to 15–30% particles larger than 45 µm (usually 200–300 $m^2$/kg Blaine). Most particles are solid spheres with average diameter 20 µm. Cenospheres and plerospheres may be present. |
| Natural materials | Besides aluminosilicate glass, natural pozzolans contain quartz, feldspar, and mica | Particles are ground to mostly under 45 µm and have rough texture |
| *Weak pozzolans* | | |
| Slowly cooled blast-furnace slag, bottom ash, boiler slag, field burnt rice husk ash | Consists essentially of crystalline silicate materials, and only a small amount of non-crystalline matter | The materials must be pulverized to very fine particle size in order to develop some pozzolanic activity. Ground particles are rough in texture |

*Source: Concrete*, 2nd edn, Mehta–Monteiro McGraw Hill, 1993, pp. 273–274, Table 8.6

Table 2.5 Admixtures and concrete durability.

| Concrete problem leading to poor durability | Probable cause of problem | Admixture that can help reduce the problem |
|---|---|---|
| Freezing and thawing | Permeable concrete. Expansion of pore water on freezing | Air-entraining agent |
| Salt scaling | Permeable concrete. Freezing-thawing damage in presence of salts | Mineral additive (e.g. microsilica) water reducer, corrosion inhibitor |
| Corrosion of reinforcement | Permeable concrete. Ingress of chloride or carbonate. Excess chloride in ingredients | Water reducer, corrosion inhibitor, high-strength additive (e.g. microsilica) |
| Alkali-aggregate reaction | Reactive aggregate, high-alkali cement | Mineral admixtures, e.g. slag, some fly ashes, microsilica |
| Chemical attack | Ingress of aggressive chemicals into permeable concrete | Selected mineral admixtures (to reduce permeability) |
| *Sulfate attack* (chemical attack involving sulfates) | Permeable concrete. Improper processing. Reaction of internal or external sulfate with cement paste components | Water-reducing admixtures. Selected mineral admixtures. Use of sulfate-resistant cement recommended |

## 2.1.4   Water

Water is a necessary component of all hydraulic concrete. It is usually used in amounts of 25–50 weight per cent of the cement. It has two engineering purposes: to enable (a) proper mixing, consolidation and finishing of the fresh mixture (workability); and (b) the chemical processes of hydration that are responsible for development and maintenance of the desired physical properties (setting, hardening, maturity). For complete hydration of a typical Portland cement, about 20–22 weight per cent of water relative to the cement content is required. Any water in access of that is theoretically needed will increase to cement paste porosity. Because of its molecular structure and chemical nature, water is an excellent solvent capable of dissolving more chemical substances than any other liquid, it exists in three phases at ambient temperatures, and is capable of penetrating even the finest pores. This makes water the most important medium also from the point of view of durability, for it is the carrier of chemical species into and out of the concrete microstructure. Without water most mechanisms of concrete deterioration could not proceed.

The following water-related items should be considered in design, production, and protection of any concrete or concrete structure: chemical nature of the water (presence of organic or inorganic components, aggressivity, alkalinity or acidity, etc.), humidity and its changes, flow rate, action of waves, and repeated drying and wetting.

## 2.2   HYDRATION OF PORTLAND CLINKER-BASED CEMENTS

*Hydration* represents a set of chemical processes between components of any cement and mixing water. The hydration reactions are strongly influenced by quality and proportions of the cementing materials used in the mix, processing procedures, and curing conditions (temperature, humidity). Hydration reactions result in the formation of new species, called *hydration products*, which give concrete the expected chemical, microstructural, and physical properties. Among properties attributable to hydration products are: setting time and workability, rate of strength development and ultimate level of strength, volume stability and creep and shrinkage and, to some degree, permeability to air and moisture, and durability.

### 2.2.1   Chemistry of hydration reactions

Considering the complexity of the anhydrous cement chemistry, it is not surprising that the products of cement hydration reactions are numerous and even more complex. The crystal structures of the products of hydration vary from perfect crystals to semi-amorphous "gels" and their specific surface areas and other surface properties also vary widely; subsequently, the hydration products differ not only in their chemical composition but also in their effect on the overall performance properties of concrete. An overview of the most important reaction products, with special focus on those of relevance to sulfate attack, is given in Table 2.6.

The chemically most active part of a concrete system is the hardened *cement paste*. It represents the cementing matrix that is responsible for such concrete properties as permeability, durability, volume stability, and mechanical strength. The cement paste is composed of:

- residual unhydrated cement components (e.g. clinker and fly ash or slag particles) and gypsum, acting as a reservoir of chemical species (and energy) needed for further reaction;
- newly-formed hydration products such as ettringite, calcium hydroxide, and calcium silicate hydrate, each of which has a function with respect to development and deterioration of concrete properties;
- porosity that to a large degree depends on the original water content of the mix and on the degree of cement hydration, and controls the migration through the concrete of chemical species responsible for concrete deterioration; and

- pore solution, the medium that fills the pores and enables (1) formation in the paste of the above cementing products; and (2) is responsible for the high alkalinity of the system.

The most important product of cement hydration is *calcium silicate hydrate* (C-S-H), sometimes referred to as calcium silicate hydrate gel (C-S-H gel). It is a nearly-amorphous, high-surface area material of variable composition, formed primarily by reaction with water of clinker components $\beta$-$C_2S$ and $C_3S$. The ratio of Ca/Si in C-S-H varies widely, a typical ratio at ambient temperature being about 1.5–1.7. Similarly, the water content of C-S-H is variable. C-S-H formed during cement hydration always contains numerous minor components, including alkalis, sulfur and alumina.

The other product of hydration of the two calcium silicates is *calcium hydroxide*, $Ca(OH)_2$ (also portlandite or CH). In the presence of cement of fly ash, slag, or microsilica, the released portlandite may react with the available silica to form additional C-S-H. For additional information on the role portlandite in hydration and deterioration of concrete consult Skalny *et al.* (2001). Both C-S-H and calcium hydroxide play important roles during sulfate attack, particularly in the presence of $MgSO_4$.

Other important products of hydration are *calcium sulfo-aluminates*: tricalcium aluminate trisulfate hydrate or ettringite (an AFt phase) and tricalcium aluminate monosulfate hydrate or monosulfate (an AFm phase). They form as the result of reactions of $C_3A$, $C_4AF$ (Fss or ferrite solid solution) or other

*Table 2.6* Hydration products: chemical and mineralogical names, oversimplified chemical formulas[#], and abbreviations.

| Compound | Formula | Abbreviation |
|---|---|---|
| Calcium hydroxide, portlandite | $Ca(OH)_2$ | CH |
| Calcium silicate hydrate | $xCaO \cdot SiO_2 \cdot yH_2O$ | C-S-H |
| Calcium sulfate dihydrate, gypsum | $CaSO_4 \cdot 2H_2O$ | $C\bar{S}H_2$ |
| Syngenite | $K_2Ca(SO_4)_2 \cdot H_2O$ | $CK\bar{S}_2H$ |
| Calcium aluminate monosulfate hydrate or monosulfate (AFm) | $Ca_4Al_2(OH)_{12} \cdot SO_4 \cdot 6H_2O$ | $C_4A\bar{S}H_{12}$ |
| Calcium aluminate trisulfate hydrate or ettringite (AFt) | $Ca_6Al_2(OH)_{12} \cdot (SO_4)_3 \cdot 26H_2O$ | $C_6A\bar{S}_3H_{32}$ |
| Thaumasite (AFt) | $Ca_3[Si(OH)_6]CO_3 \cdot SO_4 \cdot 12H_2O$ | $C_3SC\bar{S}H_{15}$ |
| Magnesium hydroxide, brucite | $Mg(OH)_2$ | MH |
| Magnesium silicate hydrate | $xMgO \cdot SiO_2 \cdot yH_2O$ | M-S-H |

[#] For more accurate and detailed information see Taylor (1997)

alumina-containing components of cement with sulfates. Under normal curing conditions, sulfates are supplied by calcium sulfate added to clinker during grinding and alkali sulfates present in the cement clinker. Calcium sulfate is interground usually in the form of gypsum or anhydrite, the purpose being the regulation of concrete setting and hardening. The four hydrated calcium sulfate compounds of primary interest during sulfate attack in concrete are gypsum (formed during sulfate attack; not gypsum added during cement processing), ettringite, monosulfate, and thaumasite (see Table 2.6) (Taylor 1997; Brown and Taylor 1999). Impotant information pertaining to these four sulfate compounds is summarized below.

## Gypsum

Calcium sulfate dihydrate, $CaSO_4 \cdot 2H_2O$ or $C\bar{S}H_2$, is a natural mineral or an industrial by-product of chemical industry. It crystallizes in the form of monoclinic tablets. Upon heating to between ca $70$–$200\,^{\circ}C$ it dehydrates to form calcium sulfate hemihydrate, $CaSO_4 \cdot 0.5H_2O$, or "soluble" anhydrite, $\gamma$-$CaSO_4$. Industrial gypsum always contains impurities, primarily fluorides and phosphates.

## Monosulfate

Monosulfate or low-sulfate calcium sulfoaluminate hydrate, $[Ca_2Al(OH)_6] \cdot SO_3$ or $C_4A\bar{S}H_{12}$, is a so-called AFm phase as it contains one (**mono**) $SO_3$ group. It crystallizes in the form of hexagonal, well-cleavaged crystals. Upon heating it decomposes in a complex series of endotherms below ca $300\,^{\circ}C$.

## Ettringite

Ettringite, $[Ca_3Al(OH)_6 \cdot 12H_2O]_2 \cdot (SO_4)_3 \cdot 2H_2O$ or $C_6A\bar{S}_3H_{32}$, also called high-sulfate calcium sulfoaluminate hydrate, is a so-called AFt phase as it contains **three** units of $(SO_4)$. It occurs as a natural mineral. It crystallizes in the form of trigonal acicular or prismatic crystals. In pure form, it decomposes above ca $85$–$90\,^{\circ}C$. A carbonate analog of ettringite, $[Ca_3Al(OH)_6 \cdot 12H_2O]_2 \cdot (CO_3)_3 \cdot 2H_2O$, has been identified.

## Thaumasite

Thaumasite, $[Ca_3Si(OH)_6 \cdot 12H_2O] \cdot (SO_4)(CO_3)$ or $C_3S\bar{S}\bar{C}H_{15}$ is an AFt phase structurally similar to ettringite, with $Al^{3+}$ substituted by $Si^{4+}$ and having one of each $(SO_4)$ and $(CO_3)$ groups within its structure. It crystallizes in the form of hexagonal acicular or prismatic crystals. Upon heating to ca $85$–$90\,^{\circ}C$ it decomposes.

Of great importance to development and deterioration of concrete engineering properties are *alkalis*. They are released into the hydrating system from the clinker components shown in Table 2.1 or from other sources such as mix water, aggregate, fly ash, admixtures, etc. In good quality concrete, alkali hydroxides, together with calcium hydroxide formed during hydration, are responsible for keeping the alkalinity (pH) of the concrete system (pore solution) at high level, of above ca 12.5, thus guaranteeing chemical stability of the hydration products and resistance of steel reinforcement to corrosion. Drastic changes in alkalinity, caused for example by excessive carbonation or some other form of chemical attack, may result in chemical and microstructural changes in the hydration products, thus leading to restructuring of the paste microstructure and subsequent deterioration of concrete mechanical properties.

*Chlorides* are present in most materials used in concrete production. Their concentration in concrete has to be limited to assure corrosion resistance of reinforcement. Excessive presence of chlorides may influence the composition of the concrete pore solution, thus affecting the form and kinetics of various degradation mechanisms.

Most field concrete has partially carbonated surface. *Carbonates*, primarily calcium carbonate (in the form of calcite or aragonite), form as a result of reaction of calcium hydroxide and, to a lesser extent, also of C-S-H present

*Table 2.7* Most important hydration reactions and reaction products of water–cement interactions[#].

| Reactant | Reaction products |
|---|---|
| Alite ($C_3S$) | CH, C-S-H |
| Belite ($\beta$-$C_2S$) | CH, C-S-H |
| $C_3A$ | Ettringite, monosulfate |
| $C_4AF$ (Fss) | Ettringite, monosulfate |
| Free lime (CaO) | CH |
| Ettringite | Monosulfate |

*Relevant reactions*

$$C_3A \quad + \quad 3C\bar{S}H_2 \quad + \quad 26H \quad = \quad C_6A\bar{S}_3H_{32}$$
$$C_3A \quad + \quad C\bar{S}H_2 \quad + \quad 10H \quad = \quad C_4A\bar{S}H_{12}$$
$$2C_3A \quad + \quad C_6A\bar{S}_3H_{32} \quad + \quad 4H \quad = \quad 3C_4A\bar{S}H_{12}$$
$$C_4ASH_{12} \quad + \quad 2C\bar{S}H_2 \quad + \quad 16H \quad = \quad C_6A\bar{S}_3H_{32}$$
$$C_3S \text{ (alite)} \quad + \quad H \quad \rightarrow \quad \text{C-S-H} + CH$$
$$\beta\text{-}C_2S \text{ (belite)} \quad + \quad H \quad \rightarrow \quad \text{C-S-H} + CH$$
$$C \quad + \quad H \quad = \quad CH$$
$$CH \quad + \quad C \quad = \quad C\bar{C}$$
$$CH \quad + \quad \bar{S} \quad + \quad H \quad = \quad C\bar{S}H_2$$

[#]For more accurate and detailed information see Taylor (1997)

in the paste with $HCO_2^-$ and $CO_3^{2-}$ ions in the ground water or atmospheric $CO_2$. As noted above, excessive carbonation may lead to decreased alkalinity of the concrete matrix, thus depassivation of the steel surface and its subsequent corrosion (Bentur *et al.* 1997; Skalny *et al.* 2001).

An overview of basic hydration reactions and reaction products formed during hydration of Portland clinker-based cements is given in Table 2.7.

It is important to recognize that concrete and the paste matrix are not inert. To the contrary, concrete is a chemically active material and its performance and deterioration in the field are highly dependent on the environmental conditions such as temperature and humidity, variations in temperature and humidity, and rate of moisture transport.

## 2.3 HYDRATED CEMENT PASTE, MORTAR AND CONCRETE

Concrete, like most other composite materials, derives its ultimate engineering properties from the nature of the original components (cement, aggregate, etc.), the mode of processing (mixing, placing, curing, etc.), the chemical and physical nature of the system after processing (hydration products, pore structure, etc.), and the environment of use (temperature, humidity, presence of aggressive chemical species, etc.). Thus, the mechanical performance and durability of concrete depend on the quality of the system as such, the primary – though not the only one – component being the cement paste. Cement paste acts as the "glue" holding the system together and its longterm stability is the primary precondition of concrete performance. Cement paste is the "heart" of concrete (Brunauer 1968).

### 2.3.1 Microstructural development

Development of the cement paste properties initiates immediately upon mixing the cementitious materials with the mixing water. It should be noted that the ultimate concrete quality is meaningfully influenced by the very early hydration of the cement components, thus proper mixing and curing procedures are crucial. This aspect of concrete production is often ignored and may lead to serious durability problems, including sulfate attack, at later ages.

Properties and durability of concrete are contingent both on the microstructure of the solids and the quality of the pore structure. The solid microstructure and the pore structure are interdependent, and are simultaneously developed in the process of hydration. Because the hydration products occupy a larger volume than the cement components from which they were formed, the original porosity, determined primarily by the mix water content, decreases as the hydration reactions proceed. During hydration the overall porosity decreases, the distribution of pore sizes changes, and the newly formed hydration products fill the voids between the remaining cement particles, previously formed hydrates, and the fine and course aggregate.

The processes of paste microstructure development and densification are most prevalent in the early days and weeks of concrete hydration; however, concrete continues to hydrate *ad infinitum* and its components react with the environment throughout its service life. Cement paste is not inert and concrete has to be treated accordingly.

The total porosity or pore volume of concrete comprises of a wide distribution of pores in the hardened cement paste, entrapped and entrained air-filled voids, and voids within the aggregate particles. In addition, under improper processing conditions, other void spaces may occur in concrete; examples are bleeding channels, plastic cracks, and honeycombing. The combined system of water- or air-filled pores and voids is referred to as *pore system*, characterized by its pore structure.

Depending on the mix design, primarily the water–cementitious materials ratio (w/cm) and processing and curing conditions, the resulting pore structure of concrete may vary widely, this variability having an important effect on the engineering performance and durability of concrete. The quality of the cement paste pore system is closely related to the permeability of the concrete, for the volume and distribution of pores controls the connectivity of the pores which, in turn, relates to diffusivity of ionic species through the concrete matrix (see Chapter 7).

The relationship between w/cm and porosity and permeability is known for decades and it is now well established that the primary factor influencing the coefficient of permeability of fully and properly cured concrete is the amount of mix water per unit of cement, i.e. the w/cm (e.g. Powers 1958; Powers *et al.* 1959; Young 1988; Garboczi and Bentz 1989; Hearn *et al.* 1994). The relationship is clearly illustrated in Figure 2.2, showing the dramatic increase in the coefficient of permeability at w/cm above approximately 0.45–0.50.

At these w/cm values, the porosity of the fully hydrated cement paste remains open, thus allowing increased diffusivity through concrete of water and chemical species dissolved in the pore water. According to Powers *et al.* (1959), the approximate ages required to produce segmentation of capillaries (to achieve discontinuity of pores) are as follows (Table 2.8).

*Table 2.8* Age required for segmentation of capillaries.

| Water–cement ratio (w/cm) | Time required to achieve discontinuity |
|---|---|
| 0.40 | 3 days |
| 0.45 | 7 days |
| 0.50 | 14 days |
| 0.60 | 6 months |
| 0.70 | 1 year |
| >0.70 | segmentation impossible |

*Source*: Powers (1959)

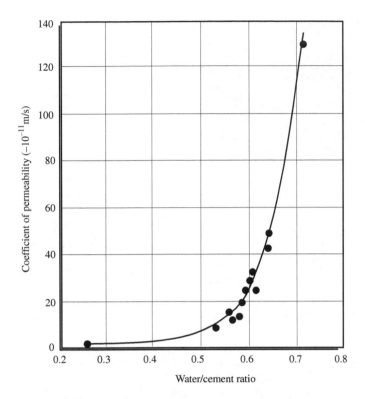

*Figure 2.2* Relationship between the coefficient of permeability and w/cm ratio for mature cement paste.

*Source*: Powers (1958). Reprinted with permission of the American Ceramic Society, PO Box 6136. Westerville, Ohio 44086-6136. Copyright 1958 by the American Ceramic Society. All right reserved

These facts are important in better understanding the importance of maintaining low w/cm and adequate curing for concrete exposed to external penetration of chemical species.

The importance of w/cm with respect to porosity is clearly demonstrated in Figures 2.3 and 2.4. Note the relative densities of concrete made with w/cm of 0.45, 0.65, and 0.85 using three microscopic techniques: UV light optical microscopy (light areas = porosity) versus SEM backscattered electron images versus SEM threshold measurements (black area = porosity; white areas = solids). It is evidently clear that higher w/cm leads to higher porosity and thus increased permeability. Highly permeable concrete is prone to deteriorate due to transport of chemicals at a higher rate.

*Microstructure* of the solids comprising the cement paste is complex indeed. As mentioned earlier, the paste is an intimate mixture of anhydrous and hydrated phases, having different chemical and mineralogical compositions,

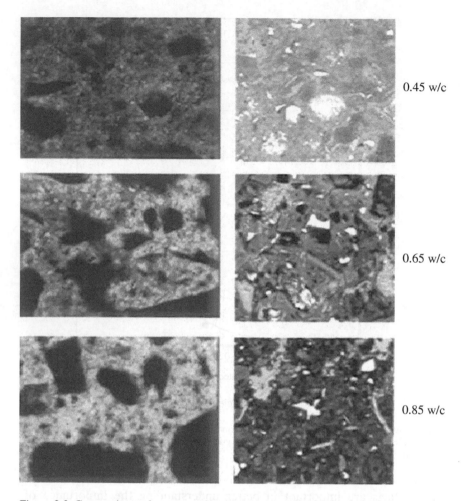

0.45 w/c

0.65 w/c

0.85 w/c

*Figure 2.3* Comparison of concrete densities for samples made with different w/cm. Light optical microscopy (florescent light) and SEM backscattered images (Photos courtesy of S. Badger).

and a wide variety of morphologies, surface areas and particle sizes. There is a time and temperature dependent exchange of chemical species between the individual solid particles and the pore solution of the concrete. The kinetics of this exchange, as explained above, depends on the permeability and diffusivity of the species through the paste matrix which, in turn, depend on the volume of mix water.

Typical cement paste and concrete morphologies are presented in Figures 2.5 and 2.6, respectively. For more details on the development of concrete microstructure we recommend to review literature sources (e.g. Scrivener

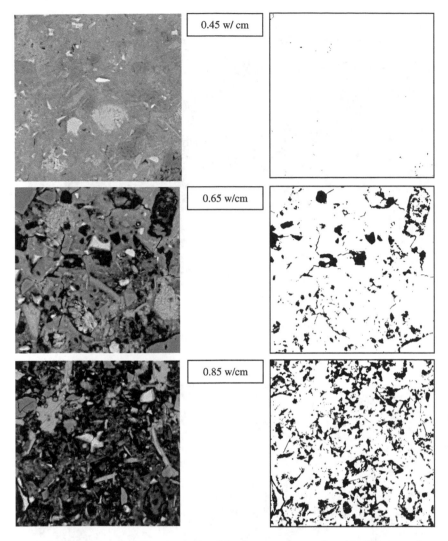

*Figure 2.4* Comparison of concrete densities for samples made with different w/cm. SEM backscattered image (BSI) and relative porosity (threshold measurements) (Photos courtesy of S. Badger).

1989; Diamond 1976). Under conditions of chemical attack, including sulfate attack, the morphology of the attacked cement paste may be modified (e.g. St John *et al.* 1998, Skalny *et al.* 1998, Diamond and Lee 1999). Such modification may, in extreme cases, lead to loss of expected properties and dramatic decrease of the expected service life.

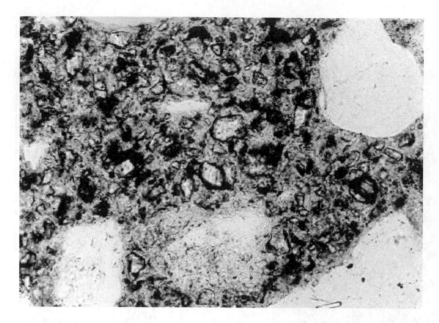

Figure 2.5 Typical microstructure of cement paste. Optical microscopy: ordinary light, field of view: 0.64 × 0.43 mm (Courtesy of U. Hjorth Jakobsen).

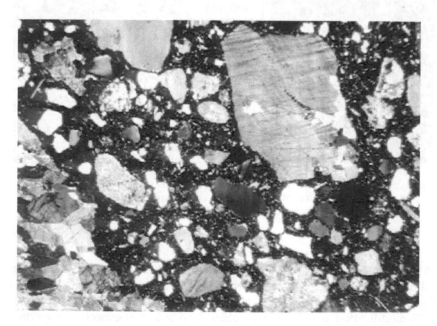

Figure 2.6 Typical microstructure of concrete. Optical microscopy: cross polarized light, field of view: 6.3 × 4.1 mm (Courtesy of G.M. Idorn Consult).

## 2.3.2   Development of physical properties

The most important property of *fresh concrete* is its workability. It can be defined as the resistivity of the fresh concrete mixture against forces of deformation, and is observed in the field as the ease of mixing, transporting (pumping), and finishing. Workability of concrete depends on the fluidity of the cement paste and the quality, amount and mixture proportions of the used aggregate. Change with time of concrete workability is usually characterized by its flowability (e.g. flow table, slump) and setting characteristics (e.g. Vicat and Proctor tests).

Like other concrete properties, the change of fresh concrete workability with time is directly influenced by the chemical reactions of hydration. As hydration progresses and more cement components transform into hydration products, the overall porosity, thus the free space between particles of the newly formed compounds, decreases; this increases the friction within the paste and between the paste and other concrete components, thus increasing the resistance to external forces. It is important to recognize that during this process of setting the concrete system may easily be damaged by excessive deformation, and this deformation may negatively influence the hardened concrete durability at later ages.

*Hardened concrete*, concrete that has set and can resist force without deformation, has to be durable – meaning it has to maintain its desired chemical and physical properties for a long time. Please note the emphasis on durability: it is a mistake to assume that strong concrete is also durable. In the majority of cases, concrete designed and processed for good durability also guarantees maintenance of concrete quality represented, typically, by volume stability and strength, whereas *strength per se is not an appropriate or adequate measure of durability*.

The most important engineering properties of hardened concrete include strength, modulus of elasticity, water tightness, and volume stability. *In toto*, the stability of these and other concrete properties in an environment constitute what is referred to as *durability*. In other words, durability is a measure of resistance of a concrete product or structure to physical and chemical "pressures." A durable structure will maintain its expected chemical, physical and engineering properties for the whole duration of its projected service life.

Chemical durability is the measure of stability of the concrete components, paste and aggregate, in the environment of its use; specifically, it is the resistance against external or internal, surface or bulk reactions that may lead to exchange of chemical species between the concrete and the environment. Such exchange may include carbonates, sulfates, chlorides, and other inorganic and organic species. The reactivity of concrete components with the above species is moisture- and temperature-dependent, thus the conditions of future use have to be considered already at the design stage and during concrete processing.

## REFERENCES

Alexander, M.G. (1998) "Role of aggregate in hardened concrete", in J. Skalny and S. Mindess (eds) *Materials Science of Concrete* V, The American Ceramic Society, Westerville, OH, pp. 119–147.

Bentur, A., Diamond, S. and Berke, N.S. (1997) *Steel Corrosion in Concrete*, E & FN Spon, London.

Brown, P.W. and Taylor, H.F.W. (1999) "The role of ettringite in external sulfate attack", in J. Marchand and J. Skalny (eds) *Materials Science of Concrete Special Issue: Sulfate Attack Mechanisms*, The American Ceramic Society, Westerville, OH, pp. 73–98.

Brunauer, S. (1968) "Tobermorite – The heart of concrete", *Scientific American* **50**: 210.

Diamond, S. (1976) In *Hydraulic Cement Pastes: Their Structure and Properties*, **2**, Slough, UK: C&CA.

Diamond, S. and Lee, R.J. (1999) "Microstructural alterations associated with sulfate attack in permeable concrete", in J. Marchand and J. Skalny (eds) *Materials Science of Concrete Special Volume on Sulfate Attack Mechanisms*, The American Ceramic Society, Westerville, OH, pp. 123–173.

Garboczi, E.J. and Bentz, D.P. (1989) "Fundamental comupter simulation models for cement-based materials", in J. Skalny and S. Mindess (eds) *Materials Science of Concrete* II, The American Ceramic Society, Westerville, OH, pp. 249–277.

Hearn, N., Hooton, R.D. and Mills, R.H. (1994) "Pore structure and permeability", in J. Lamond and P. Klieger (eds) *Tests and Properties of Concrete*, pp. 240–262.

Lawrence, C.D. (1995) "Delayed ettringite formation: A problem?", in J. Skalny and S. Mindess (eds) *Materials Science of Concrete* IV, The American Ceramic Society, Westerville, OH, pp. 113–154.

Mehta, P.K. and Monteiro, P.J.M. (1993) *Concrete – Microstructure, Properties and Materials*, 2nd edn, The McGraw-Hill Companies, Inc., New York.

Powers, T.C. (1958) "Structure and physical properties of hardened Portland cement paste", *Journal of the American Ceramic Society* **41**: 1–6.

Powers, T.C., Copeland, L.E. and Mann, H.M. (1959) "Capillary continuity or discontinuity in cement pastes", *J. of Portland Cement Association R&D Laboratories* **1**(2): 38–48.

Scrivener, K.S. (1989) "Microstructure of concrete", in J. Skalny (ed.) *Materials Science of Concrete* I, The American Ceramic Society, Westerville, OH, pp. 127–162.

Skalny, J., Diamond, S. and Lee, R.J. (1998) "Sulfate attack, interfaces and concrete deterioration", in A. Katz, A. Bentur, M. Alexander and G. Arliguie (eds) *Proceedings, RILEM 2nd International Conference on The Interfacial Transition Zone in Cementitious Composites*, NBRI Technion, Haifa, pp. 141–151.

Skalny, J., Gebauer, J. and Odler, I. (eds) (2001) *Materials Science of Concrete Special Volume: Calcium Hydroxide in Concrete*, The American Ceramic Society, Westerville, OH.

St John, D.A., Poole, A.B. and Simms, I. (1998) *Concrete Petrography*, London: Arnold.

Taylor, H.F.W. (1997) *Cement Chemistry*, 2nd edn, Thomas Telford Publishing, London.

Young, J.F. (1988) "A review of pore structure of cement paste and concrete and its influence on permeability", in D. Whiting (ed.) *Proceedings, Permeability of Concrete*, SP 108, American Concrete Institute, Detroit, MI, pp. 1–18.

# 3 Concrete deterioration

## 3.1 PRINCIPAL CAUSES OF CONCRETE DETERIORATION

Deterioration of concrete may be caused by chemical and physical processes, or their combination. Chemical mechanisms include leaching of paste components, carbonation of calcium hydroxide and C-S-H, paste deterioration by exposure to aggressive chemicals (acids, agricultural chemicals, sulfates), corrosion of steel reinforcement, and alkali-aggregate reactions. Physical or mechanical causes of concrete deterioration are represented by abrasion, erosion, cavitation and, most important, by freezing and thawing cycles. Most chemical mechanisms of deterioration involve damage to the cement paste matrix, but deterioration of the paste-aggregate interface, the aggregate itself, or the pore structure often accompanies the paste deterioration. Sulfate attack, the subject of this monograph, is a form of chemical mechanism.

A common denominator of most mechanisms of deterioration is access to concrete of moisture. Without water, with mechanical damage such as abrasion being an exception, no mechanism of deterioration can proceed. Examples of such processes are of both chemical (e.g. alkali–silica reaction, corrosion of reinforcement, sulfate attack) and physical (e.g. freezing–thawing) nature.

The general factors that accelerate or retard chemical damage of concrete are summarized by ACI as follows (Table 3.1).

When examining concrete deterioration, one has to take into consideration the local microclimate the concrete is exposed to. The concrete surface durability may depend on the local temperature and humidity, thus an important consideration is exposure or lack of exposure to rain, direct sun, etc.

It should be remembered that field concrete, in contrast to laboratory-made specimens, is usually exposed to more than one mechanism of deterioration; such situation may lead to synergistic interactions resulting in increased rate of deterioration. If combined mechanisms are operative, it may be often difficult to clearly identify the primary cause of the problem. Examples of combined mechanisms of concrete deterioration are acid attack and leaching, alkali–silica reaction and internal sulfate attack (e.g. delayed ettringite formation), and carbonation and reinforcement corrosion.

*Table 3.1* Factors influencing chemical attack on concrete.

| *Factors which accelerate or aggravate attack* | *Factors which mitigate or delay attack* |
| --- | --- |
| (1) High porosity due to<br>  i high water absorption<br>  ii permeability<br>  iii voids | (1) Dense concrete achieved by<br>  i proper mixture proportioning<br>  ii reduced unit water content<br>  iii increased cementitious material content<br>  iv air entrainment<br>  v adequate consolidation<br>  vi effective curing |
| (2) Cracks and separations due to<br>  i stress concentrations<br>  ii thermal shock | (2) Reduced tensile stress in concrete by<br>  i using tensile reinforcement of adequate size, correctly located<br>  ii inclusion of pozzolan (to suppress temperature rise)<br>  iii provision of adequate construction joints |
| (3) Leaching and liquid penetration due to<br>  i flowing liquid<br>  ii ponding<br>  iii hydraulic pressure | (3) Structural design<br>  i to minimize areas of contact and turbulence<br>  ii provision of membranes and protective-barrier system(s) to reduce penetration |

*Source*: *Guide to Durable Concrete* (ACI 201-2R-92). Reprinted with permission by the American Concrete Institute

### 3.1.1    Deterioration caused by dissolution of paste components

The most common phenomenon in this category is the dissolution of the cement paste constituents, in particular portlandite, $Ca(OH)_2$, present in cement paste primarily as a consequence of hydration of clinker calcium silicates. $Ca(OH)_2$ can be removed from the paste matrix simply due to its inherent high solubility at a high permeability of the concrete when placed in an environment with flowing water, or it can be the result of chemical reactions within the paste and the subsequent removal of calcium from the system (e.g. Marchand and Gerard 1995). The latter scenario is closely related to issues discussed briefly in the following section and in more detail throughout the book.

### 3.1.2    Deterioration caused by ingress of external chemicals

Ingress of external species into concrete and the subsequent removal of the reaction products from the system are clearly porosity and permeability dependent. In the absence of a permeable pore structure or, better, if the rate of penetration through the concrete matrix pore structure is low, the probability of deterioration is dramatically decreased.

The rate of deterioration of a concrete structure by external chemicals will depend on the following conditions:

- concentration, chemical identity, and solubility in water of the external reactant in the soil or water which is in contact with concrete;
- presence of water in the soil in contact with the structure and its mobility;
- concrete quality, including density (w/cm, porosity), degree of hydration, and the resulting permeability; type of cement used; absence of plastic cracks, etc.;
- atmospheric environment of use, including temperature and temperature variations, humidity and humidity variations, cycles of drying and wetting (effect of wind conditions), etc.

For all types of external chemicals, whether organic or inorganic in nature, the probability of deterioration is minimized if the concrete has low permeability and is placed in a dry environment at constant temperature and humidity. Understandably, such conditions are uncommon in most applications. It is important, therefore, to recognize the above facts and consider right at the design phase materials and processing conditions that will minimize the potential damage. Recognition of potential problems is especially important for construction in geographical areas known to have soils containing aggressive components, and in some industrial and agricultural applications. See for example (e.g. DePuy 1994; Reinhardt 1997; Marchand *et al.* 1998).

Examples of species known to damage concrete are as follows: soft waters, acidic waters, sulfates in ground water, agricultural chemicals, etc. The conditions listed above apply also to penetration of chlorides; however, chloride penetration affects primarily the chemical stability (corrosion resistance) of the reinforcing steel rather then the paste itself.

### 3.1.3   Expansive reactions with aggregate

The best-known phenomena in this category are deterioration mechanisms of concrete referred to as alkali-aggregate reactions. The most prevalent alkali-aggregate reaction is alkali–silica reaction, ASR, characterized by the reaction of $OH^-$ ions with silica contained in reactive amorphous silicates in aggregate. The reaction leads to breaking of the -Si-O-Si- bonds in the silicate structure and formation of calcium–alkali–silicate gel of variable composition that may, under certain conditions, lead to cracking of the aggregate, expansion of the cement paste matrix and, ultimately, to cracking and complete destruction of the concrete structure. A typical example of ASR damage is shown in Figure 3.1. For more information, see selected specialized literature (e.g. ASR 1974–2000; Diamond 1989; Helmuth and Stark 1992; SHRP 1993).

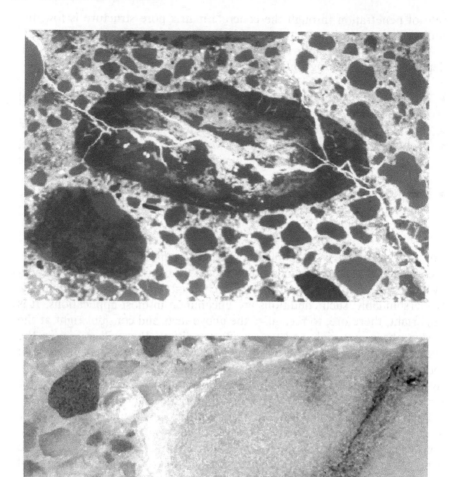

*Figure 3.1* (a) Alkali silica reactive flint aggregate particle in a concrete bridge exposed to sea water. Field of view: 6.3×4.1 mm. Light optical microscopy: fluorescent light (Photo courtesy of U. Hjorth Jakobsen); (b) Alkali–silica reactive particle in concrete made with aggregate containing both flint and silicious grains in aggregate. Field width: 5 mm, light optical microscopy (Photo courtesy of P. Stutzman).

It should be noted that field experience shows ASR to often occur in combination with the DEF-type of internal sulfate attack. More detailed information on this phenomenon is given in Chapters 4 and 8.

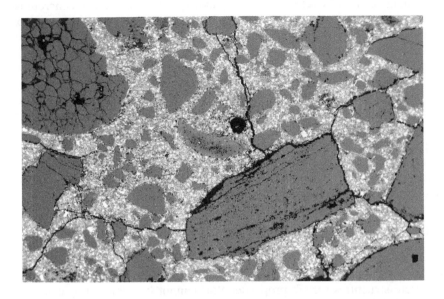

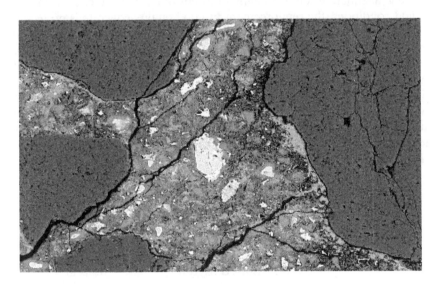

*Figure 3.2* Microstructural damage to concrete aggregate and paste exposed to freezing–thawing (Photos courtesy of P. Stutzman).

### 3.1.4  Frost-related deterioration

Together with damage caused by the corrosion of reinforcing steel, frost-related
decay is the most common cause of concrete deterioration. Such type of deterioration is prevalent primarily in geographical areas with frequent fluctuation
of temperature around the freezing temperature of water (see Figure 3.2).

Damage caused by repeated freezing and thawing of water in concrete is physical in nature. It is caused primarily by deterioration of the paste matrix due to repeated mechanical stresses resulting from volume changes associated with freezing of water into ice, but aggregate may also be affected (D-cracking). It has to be recognized, however, that the chemical composition of the water in concrete (pore solution) does have an effect on the freezing process. Protection against this mode of deterioration includes use of high density, good quality concrete allowing only low water adsorption, use of air-entraining admixtures, and limiting the access of water to the structure. Use of air-entraining admixtures is the common methodology of protection against repeated freezing and thawing, but by itself it is not a solution unless used with good quality concrete. For more information, see available literature (e.g. Pigeon and Pleau 1994; Marchand *et al.* 1995).

### 3.1.5   Corrosion of embedded steel

The vast majority of concrete structures, whether plain or prestressed, are erected with imbedded reinforcing steel. The function of the steel is to give concrete structures certain properties that cannot be achieved by concrete itself, e.g. adequate flexural strength, modulus of elasticity, etc. Under regular field conditions, the reinforcement in good quality concrete is protected from aggressive corroding environment by the alkaline environment of the concrete cover and by use of otherwise protected steel surface, or both. The steel surface itself is covered by an oxide layer that is stable in alkaline, but not under more acidic conditions.

Under field conditions, however, the ideal conditions are often not met. In lower quality concrete, for example concrete made with excessive water and thus posessing high porosity and permeability, the reinforcing steel may undergo oxidation resulting in electrochemical (cathodic plus anodic) removal of Fe atoms from the surface.

Anodic reaction:   $2Fe^0 - 4$ electrons $\rightarrow 2Fe^{2+}$
Dissolution of Fe atoms — Ions in solution

Cathodic reaction:   $O^{2-} + 2H_2O + 4$ electrons $\rightarrow 4(OH^-)$
Dissolved oxygen molecules — Ions in solution

Under such conditions the protective iron oxide layer may get damaged, leading to excessive formation of one of several varieties of rust, followed by expansion and delamination of concrete (see Figure 3.3).

High concrete porosity and high concentration of cracks reaching the depth of reinforcement increase also the rate of carbon dioxide ingress, this

*Figure 3.3* Concrete overpass damaged by severe corrosion of reinforcing steel (Photo courtesy of N. Berke).

leading to carbonation of the calcium hydroxide, $Ca(OH)_2$, present in the cement paste, to form calcium carbonate, $CaCO_3$.

$$Ca(OH)_2 + CO_2 \rightarrow CaCO_3 + H_2O$$

This may result in a drop of pH in the cement paste that is in immediate contact with the steel surface, thus allowing the corrosion of steel to take place.

Corrosion of reinforcement steel may be accelerated under conditions where other mechanisms of damage are in action simultaneously. For example, simultaneous freeze–thaw conditions or exposure of concrete to aggressive sulfates may accelerate the rate of corrosion. In a similar manner, chemical deterioration of concrete may be aided by conditions leading to reinforcement corrosion, such as access of chlorides and sulfates. For additional information, we recommend specialized literature (e.g Uhlig 1971; Bloomfield 1996; ACI 1992; Bentur *et al.* 1997).

### 3.1.6  Abrasion, erosion and cavitation

Forms of mechanical deterioration include abrasion, erosion, and cavitation, all damage mechanisms caused by frictional interaction of gases, fluids or solid particles with concrete surface.

*Abrasion* is characterized by wear of concrete surface by repeated friction caused, for example, by continuous driving on concrete pavements or industrial floors.

*Erosion* is a special case of abrasion, involving wear of concrete surface by water and wind-born or water-suspended particles.

*Cavitation* occurs in situations where sudden change in velocity or direction of water in contact with concrete surface causes formation of a zone of sub-atmospheric pressure, allowing formation and subsequent collapse of pockets of vapor. The collapse or implosion of these low-pressure pockets results in localized high-pressure impact on the concrete structure, leading to mechanical damage. (see e.g. Mindess and Young 1981; Mehta and Monteiro 1993; Neville 1997).

As is the case with other mechanisms of deterioration, the rate of damage due to abrasion, erosion or cavitation may accelerate if other chemical or physical deterioration mechanisms are operative at the same time.

## 3.2   SELECTION OF MATERIALS

The selection of concrete mix components should be done by considering the:

- desired properties of concrete in the environment of the expected use;
- quality of components needed to achieve the desired concrete properties, and;
- availability and economic feasibility of the available sources.

Concrete components have to be selected keeping in mind the expected service life of the structure in the environment of use. This is generally done by compliance with the best practices of concrete making and by adhering to national and local standards, codes and regulations. Unfortunately, for various reasons, such compliance is not always the case, therefore use of common sense and conservative but economically feasible approach to structural design is recommended.

### 3.2.1   Importance of mix design

To produce concrete possessing the expected mechanical and chemical properties and long-term durability, the mixture design should:

- allow for production of fresh concrete that can be properly homogenized, placed, and finished and
- give the maximum possible packing density to minimize ingress of chemical species while guaranteeing the desired structural integrity.

This is even more important when the structure is to be erected in a hostile environment, such as cold climates or chemically aggressive soil, or when availability of quality concrete raw materials is in question. It is advisable, as the first approximation, to always use as low as possible w/cm, because practically all mechanisms of concrete deterioration are water sensitive. Decreasing the access of water to a structure and minimizing transport of chemical species through concrete are thus crucial.

## 3.3 CONCRETE PROCESSING

Even a properly designed concrete mixture may fail prematurely if development of the desired chemical, microstructural, and mechanical properties is compromised by inadequate or erratic processing. We dare to say that only a very small volume of bad concrete produced worldwide is related to causes other than inadequate processing. Damage to concrete is rarely caused entirely by low-quality concrete components, as in almost all cases improper processing is a contributing factor.

### 3.3.1 Mixing, curing, placing, finishing, and maintenance

Assuming proper structural and materials design, most processing damage is introduced during mixing and curing. Placement techniques are not immune to mistakes either and concrete maintenance is routinely ignored. Fortunately, concrete is a somewhat forgiving material and, in spite of common abuse most, but not all, concrete performs its function adequately.

Proper processing is particularly important when concrete is placed into an aggressive environment. Inadequate mixing, for example, may lead to inhomogeneous hardened concrete matrix, both in the solid matter itself and in distribution of the air voids and, consequently, may result in increased porosity and/or internal stresses. Such concrete is easier attacked by chemical species from the environment.

Inadequate curing is a well-known reason for durability problems. In at least one type of sulfate attack, namely so-called delayed ettringite formation (DEF), improper curing practices are the most probable cause of damage. Learn more about this form of internal sulfate attack in a special section of this monograph (Section 4.3.2).

Placing and finishing of concrete, especially in "low-technology" applications like residential housing and street curbs, are most commonly abused. Not unlike other problems in concrete processing, the reason is the inadequate awareness and utilization of existing knowledge. Concrete is often considered to be an "artificial stone" that can be abused without penalty. This is not the case. Concrete properties, including durability and service life, are preconditioned by proper processing and timely and repeated maintenance.

## 3.4   EFFECT OF ENVIRONMENTAL EXPOSURE

Most concrete is stable in the environment of its use. Such concrete is designed and produced to have dense, impermeable, and mechanically sound macro- and microstructure, the components of which are chemically stable. If these conditions are violated, concrete, like all chemically active materials, will react with its environment to produce chemical species that are unstable and whose formation may result in microstructural changes that could severely compromise the expected concrete properties.

### 3.4.1   Effect of chemical environment

Knowledge of the chemical nature of the soil and ground water to which a concrete structure will be exposed during its service is crucial. The chemical nature of the environment should be recognized *before* the concrete is designed for the particular environment. As will be shown elsewhere in more detail, such knowledge is most important in geographical areas with high sulfate concentrations, and it is not only the concentration of sulfate anions *per se* but also the type and concentration of the accompanying cations that should be taken into consideration.

### 3.4.2   Effects of temperature and humidity changes

Chemical reactions of cement hydration and concrete deterioration, like most other chemical reactions, are influenced by the reaction temperature and, to a lesser extent, by the relative humidity. With a few exceptions, most chemical reactions are accelerated by increased temperature, although different chemical species are influenced to different degrees. As a result, increased or decreased temperatures change the relative reactivities of the species involved in hydration and deterioration processes, leading to unexpected modifications in solids density, microstructure, and chemical equilibria when compared to the usual, expected conditions.

It is often assumed that concrete processing can be successful in a wide temperature range without a penalty. Although this may be correct in most cases, one should not take this for granted. Under unfavorable combination of processing conditions (e.g. placing a concrete mixture with high content of high-surface area cement at high ambient temperature and low relative humidity), the resulting concrete may be susceptible to damage. Such damage may be the result of ingress of water and of dissolved ionic species due to high concentration in the concrete matrix of thermally-induced cracks, inadequate degree of hydration, and resulting low early strength.

Curing of concrete is a thermo-chemical process. To properly cure concrete to give the best expected performance, a knowledgeable compromise has to be used that enables all important components of the cement paste to perform their designed function in the given time. That is the reason why not only the maximum temperature and time of curing are important, but also the time of precuring, rate of heating and cooling, and the relative humidity of the curing environment have to be taken into consideration.

It is important to note that the individual hydration and deterioration reactions of any concrete system are closely interrelated and influence each other. This complexity is the reason why a particular damage mechanism may enable another destructive mechanism to occur simultaneously; a relevant example is the often-observed simultaneous occurrence of alkali–silica gel (ASR) and internal sulfate damage (DEF).

## 3.5  KNOWN PREVENTION TECHNIQUES

Prevention of concrete damage by any mechanism of deterioration will depend on at least the following:

* awareness of the available knowledge on the chemical and physical processes that control the concrete quality in the given environment of use;
* consideration of the available knowledge beginning with the design stage (both concrete mix design and structural design), through processing, to maintenance;
* proper use of the existing knowledge in development of standards and test methods; and
* adherence to the standards, codes, and other regulations.

Literature on how to prevent *sulfate attack* in concrete, whether caused by internal or external sulfate sources, is readily available. Although the recommendations given in the technical literature do reflect on the complexity of the underlying scientific phenomena and the lack of complete mechanistic understanding of all related issues, they are based on simple, common sense, well-publicized ideas. It remains a fact that, although standards and testing methods are not perfect and do not always take into consideration of all the available knowledge (see Chapter 9 on standardization), adherence to standards and building codes can prevent occurrence of sulfate attack in most cases.

In the Appendix part of this monograph, find recommendations for prevention of various forms of sulfate attack. The recommendations are based on literature references reflecting the recent best practices of concrete making (Deutscher Ausschuss fur Stahlbeton 1989; CEN 1998; Hobbs 1998; Skalny and Locher 1999, etc.). For detailed justification of the proposed preventive measures see the mechanistic and practical data discussed throughout the book.

## REFERENCES

ACI (1992) ACI Committee 201, "Corrosion of steel and other materials embedded in concrete", Chapter IV, In *Guide to Durable Concrete*, Manual of Concrete Practice, Part 1, The American Concrete Institute, Detroit.

ASR (1974–2000) In *Proceedings of the International Conferences on Alkali-Aggregate Reaction* held in Koge (1974), Cape Town (1981), Kyoto (1989), Melbourne (1996), and Quebec (2000).

Bentur, A., Diamond, S. and Berke, N.S. (1997) *Steel Corrosion in Concrete*, E & FN Spon, London.

Bloomfield, J.P. (1996) *Corrosion of Steel in Concrete: Understanding, Investigation and Repair*, E & FN Spon, London.

CEN (1998) CEN/TC 104/SC 1 N 308, "Common rules for precast concrete products (Draft 04/98)", September.

DePuy, G.W. (1994) "Chemical resistance of concrete", in Lamond and Klieger (eds) *Tests and Properties of Concrete*, STP 169C, ASTM, Philadelphia, 263.

Deutscher Ausschuss fur Stahlbeton (1989) *Recommendation on the Heat Treatment of Concrete* (in German), September.

Diamond, S. (1989) "ASR – another look at mechanisms", in K. Okada *et al.* (eds) *Proceedings of the 8th International Conference on Alkali-Aggregate Reaction*, The Society of Materials Science, Tokyo.

Helmuth, R. and Stark, D. (1992) "Alkali-silica reactivity mechanisms", in J. Skalny (ed.) *Materials Science of Concrete*, Vol. III, The American Ceramic Society, Westerville, OH, pp. 131–208.

Hobbs, D.W. (1998) *Minimum Requirements for Durable Concrete: Carbonation- and Chloride-induced Corrosion, Freeze–thaw Attack and Chemical Attack*, British Cement Association.

Marchand, J. and Gerard, B. (1995) "New developments in the modeling of mass transport processes in cement-based composites", in V.M. Malhotra (ed.) *ACI Special Publication SP-154*, American Concrete Institute, Detroit, MI, pp. 169–210.

Marchand, J., Pleau, R. and Gagne, R. (1995) "Deterioration of concrete due to freezing and thawing", in J. Skalny and S. Mindess (eds) *Materials Science of Concrete*, vol. IV, The American Ceramic Society, Westerville, OH, pp. 307–400

Marchand, J., Gerard, B. and Delagrave, A. (1998) "Ion transport mechanisms in cement-based materials", in J. Skalny and S. Mindess (eds) *Materials Science of Concrete*, vol. V, The American Ceramic Society, Westerville, OH, pp. 283–354.

Mehta, P.K. and Monteiro, P.J.M. (1993) *Concrete: Structure, Properties, and Materials*, McGraw-Hill, New York.

Mindess, S. and Young, J.F. (1981) *Concrete*, Prentice-Hall, Englewood Cliffs, New Jersey.

Neville, A. (1997) *Properties of Concrete*, 4th edn, John Wiley and Sons, New York.

Pigeon, M. and Pleau, R. (1994) *Durability of Concrete in Cold Climates*, Chapman & Hall, London.

Reinhardt, H.W. (ed.) (1997) *Penetration and Permeability of Concrete: Barriers to Organic and Contaminating Liquids*, RILEM Report 16, E & FN Spon, London, 332 pp.

SHRP (1993) *Alkali–Silica Reactivity: An Overview of Research*, SHRP-C-342, Washington, D.C.

Skalny, J. and Locher, F. (1999) "Curing practices and delayed ettringite formation – the European experience", *Cement, Concrete, Aggregate*, ASTM, June.

Uhlig, H.H. (1971) *Corrosion and Corrosion Control*, John Wiley and Sons.

# 4    Sulfate attack

## 4.1    FORMS OF SULFATE ATTACK

*Sulfate attack* is the term used to describe a series of chemical reactions between sulfate ions and the components of hardened concrete, principally the cement paste, caused by exposure of concrete to sulfates and moisture. As is the case with other aggressive chemicals, sulfates are potentially most deleterious to concrete when present in gaseous or liquid form, the latter situation being the most common; attack by solid sulfate-containing chemicals is rare.

First we would like to emphasize that the chemistry of sulfate attack is independent of the source of sulfate; as will be shown later, the processes are governed by the same physico-chemical principles. The differences in the consequences of any sulfate attack are caused by the environmental and physical conditions under which these reactions proceed.

The chemistry of sulfate attack is complex and involves numerous overlapping reactions. Because of this complexity, one of the problems encountered in the relevant literature on concrete durability is the question of *definition* of sulfate attack. Is attack by sulfuric acid considered to be sulfate attack? Can damage caused by formation and subsequent repeated recrystallization of a hydrous sulfate salt into an anhydrous one be considered the result of sulfate attack? Is the reaction mechanism of $MgSO_4$ originating from sea water different from the mechanism by which $MgSO_4$ from ground water is attacking concrete? Are there any reasons to separate the so-called *chemical sulfate attack* from so-called *physical sulfate attack* or *acid attack*? Is the so-called *delayed ettringite formation* (DEF) a special form of *internal* sulfate attack? Is DEF expansion caused by the same mechanism as the *classical* sulfate attack?

According to ACI's Guide to Durable Concrete (ACI 1992), there are two mechanisms that can be considered to be *sulfate attack*: formation of gypsum and formation of ettringite. Both of these reaction products are believed to damage concrete by increases in overall solid volume. In some literature, these forms of attack are referred to as *gypsum corrosion* and *sulfoaluminate corrosion*, respectively. According to Mehta and Monteiro (1993), the two manifestations of *sulfate attack* are *expansion* (caused primarily by formation of ettringite, possibly of gypsum) and *progressive loss of strength and loss of mass* (caused by deterioration of the cohesiveness of the cement hydration products). Literature sources also show that, depending on the concentration and form of sulfates, the local environmental conditions and processing practices, the formations of gypsum and ettringite do not have to be necessarily expansive, and damage to concrete by mechanisms other than solid volume increase may become operative (e.g. Taylor 1997; Brown and Taylor 1999; Mehta 2000). This fact is well described by Taylor (1997):

> Ground water containing sulfate is often also high in magnesium, and sulfate attack has often been discussed in terms of reaction between solid phases in cement paste and dissolved compounds, such as $Na_2SO_4$ or $MgSO_4$, in the attacking solution. This obscures the fact that the reactions of the cations and anions in that solution are essentially separate. With $NaSO_4$, the reaction is one of the $SO_4^{2-}$ ions, but with $MgSO_4$ the $Mg^{2-}$ also participates.

As a consequence, a solution of $NaSO_4$ may cause both sulfate attack and ASR, and a solution of $MgSO_4$ may cause sulfate attack and reactions forming brucite or other Mg-containing species. These and additional facts show that the relevant reaction mechanisms are numerous and complex. In the following sections, these issues will be reviewed in more detail.

In numerous literature sources, sulfate attack is categorized into *chemical* versus *physical* and into *internal* versus *external*. *Chemical* sulfate attack is considered to be the result of *chemical* reactions involving sulfate anion, $SO_4^{2-}$. Example of such reaction is formation of ettringite from monosulfate and gypsum, according to the following stoichiometric reaction:

$$C_4A\bar{S}H_{12} \; + \; 2C\bar{S}H_2 \; + \; 16H \Rightarrow C_6A\bar{S}_3H_{32}$$

Monosulfate    Gypsum       Water    Ettringite

where C is CaO, A is $Al_2O_3$, $\bar{S}$ is $SO_3$, and H is $H_2O$. The reaction is known to result in an increase in solid volume of the system, and may or may not lead to the expansion of concrete.

The so-called *physical* sulfate attack, often called also sulfate *salt crystallization* or *salt hydration distress*, usually refers to (a) formation from the solution of sodium sulfate decahydrate, $Na_2SO_4 \cdot 10H_2O$, followed by (b) its repeated recrystallization into sodium sulfate anhydrite, $Na_2SO_4$, and vice versa:

$$2Na^+ \; + \; SO_4^{2-} \quad \Rightarrow \quad Na_2SO_4 \cdot 10H_2O \;\; (a)$$

(solution)    (evaporation)    (solid)

$$Na_2SO_4 \cdot 10H_2O \quad \Leftrightarrow \quad Na_2SO_4 \qquad (b)$$

Mirabilite    (repeated    Thenardite

recrystallization)

This temperature-dependent process leads to repeated increase in volume thus, if the process happens within the concrete matrix, it will lead to fatigue of the cement paste and its subsequent loss of cohesion. The repeated expansion–contraction is the origin of the term *physical* in describing this form of sulfate attack. The term *physical* does not seem to be correct in characterizing this reaction, as hydration and dehydration of sodium sulfate, not unlike portlandite or ettringite or gypsum formation, are *chemical* processes – *chemical processes with physical consequences*. As will be discussed later, the physical and chemical aspects of action of sulfate on concrete cannot be clearly separated (see Section 4.11).

*Internal* sulfate attack refers to situations where the source of sulfate is internal to concrete. The source of sulfate can be the cement, supplementary materials such as fly ash or slag, the aggregate, the chemical admixtures, or the water. Two examples of such internal sulfate attack are the classical attack by excess (with respect to the clinker aluminate phase) of cement sulfate and the so-called *delayed ettringite formation* (DEF).

*External* sulfate attack is caused by a source external to concrete. Such sources include sulfates from ground water, soil, solid industrial waste, and fertilizers, or from atmospheric $SO_3$, or from liquid industrial wastes.

We will avoid the artificial distinction between chemical and physical sulfate attacks. First of all, as stated previously, the term *physical* attack is used incorrectly. But, most important, as we will show, sulfate attack is a complex set of overlapping *chemical* and *physical* processes, the chemical and mechanical consequences of which depend not only on the concentration, form, and solubility of the chemical species involved – by themselves very complex – but also on the concrete processing practices, the overall quality of concrete, and the environment in which the concrete is made and used.

### 4.1.1 Manifestations of sulfate attack

Some of the *visible* examples of damage caused by reaction of concrete components with sulfates have been given in Figures 1.1–1.4. Such visible damage includes spalling, delamination, macrocracking and, possibly, loss of cohesion. All of these phenomena are consequences of the chemical processes *invisible* to naked eye, including adsorption–desorption phenomena, dissolution–precipitation of colloidal and crystalline phases, recrystallization, etc.

Some or all of the following complex physico-chemical processes may be involved in a typical sulfate attack (e.g. Skalny *et al.* 1998; Diamond and Lee 1999; Glasser 1999; Diamond 2000; Brown and Doerr 2000):

*Dissolution or removal from the cement paste of calcium hydroxide.* The consequence of such dissolution or removal is leaching of calcium and hydroxyl ions out of the system, thus a possible decrease in the alkalinity (pH) of the paste, or sulfonation of the $Ca^{2+}$ ions to form potentially expansive compounds such as ettringite or gypsum.

*Complex and continuous changes in the ionic composition of the pore liquid phase.*   Composition of the liquid changes continuously even under normal processes of hydration, especially in early ages; however, under chemical attack and/or changes in the environmental conditions, these changes may lead to formation of conditions allowing deterioration of the paste matrix. An example of such change is penetration into the surface pore structure of $CO_2$: carbonation leads to the removal from the liquid phase of $Ca^{2+}$ and $OH^-$ ions, decrease in pH, and subsequent instability of the ettringite in the carbonated zone.

$$Ca^{2+} + 2OH^- + CO_2 \rightarrow CaCO_3 + H_2O$$

*Adsorption or chemisorption of ionic components present in the pore liquid phase on the surface of the hydrated solids present in the cementing system.* Specifically, as an example, the sulfate and aluminate ions may be sorbed by the C-S-H under some conditions, and then released when the conditions change. As will be discussed later, this mechanism seems to be of crucial consequence in delayed formation of ettringite.

*Decomposition of still unhydrated clinker components.*   There is some evidence to show that early penetration of sulfates into the system may lead to or promote decomposition or, better, decalcification of the anhydrous calcium silicates present in the clinker. The consequences of such process are increase in porosity, and formation of hydrous silica or Mg-silicate hydrate-compounds that do not possess cementing properties (more about this phenomenon later), or both. Under some conditions, the ferrite phase, usually not considered to be involved in sulfate attack reactions, may become more reactive, thus leading to release of aluminate and ferrite ions that can subsequently participate in formation of ettringite.

*Decomposition of previously formed hydration products.*   An example is decalcification of calcium hydroxide, $Ca(OH)_2$, and even decomposition of C-S-H, calcium silicate hydrate, the principle binder in the cement system. Such partial or even complete destruction ultimately leads to loss of cementing properties (paste cohesion, paste-to-aggregate bond, strength, and durability).

*Formation of gypsum.*   Its formation is believed to be associated with a limited increase in volume and becomes an issue at sulfate concentrations above ca 3,000 ppm. However, its formation uses up the available sources of calcium, thus gypsum formation may have secondary consequences by promoting other sulfate-related phenomena.

*Formation of ettringite.*   Due to its low solubility, ettringite can form at relatively low sulfate concentrations; this is the reason why ettringite-triggered expansion is more prevalent than gypsum-generated expansion.

*Formation of thaumasite.*   Thaumasite is believed to form primarily at temperatures below ca 5–10 °C; however, evidence is mounting that it may form even at elevated temperatures. Damage due to thaumasite is reported to possibly be more disruptive than that due to ettringite, as it is formed from C-S-H.

*Formation of brucite and magnesium silicate hydrate.*   In the presence of $Mg^{2+}$ cations, the sulfate attack reactions are more damaging than in the presence of calcium or sodium cations, in that magnesia itself leads to deleterious processes supplementing those caused by formation of sulfoaluminates. Both reactions occur simultaneously and the damage to concrete becomes multifaceted. Replacement of Ca in C-S-H by Mg leads to loss of cementing properties. Formation of brucite or Mg-silicates is an indication of severe Mg-sulfate attack.

*Formation of hydrous silica (silica gel).*   This is another indication of severe sulfate attack, primarily in the presence of Mg-sulfate. Please note that silica gel may be present also as the consequence of severe carbonation.

*Penetration into concrete of sulfate anions and subsequent formation and repeated recrystallization of sulfate salts.*   As highlighted above, recrystallization of thenardite to/from mirabilite is an example of this phenomenon. However, under certain circumstances formation of other salts has been observed: $CaCO_3$ (calcium carbonate), $Na_3H(CO_3)_2 \cdot 2H_2O$ (trona), $8.5CaSiO_3 \cdot 8.5CaCO_3 \cdot CaSO_4 \cdot 15H_2O$ (birunite), $NaHCO_3$ (nahcolite), NaCl (halite), among others. As will be discussed later, the presence of these compounds may be, under some conditions, indicative of or related to sulfate attack.

High initial permeability of the cement paste or damage due to other simultaneous mechanisms increase atmospheric carbonation and penetration of external chemical species present in the local environment, or both. Under such conditions, some of the above reactions may be accelerated or decelerated, thus increasing the overall complexity of the attack. For example, it is possible to have severe deterioration of concrete due to sulfate attack in the absence of extensive volumetric expansion.

## 4.2   SOURCE OF SULFATES IN SULFATE ATTACK

### 4.2.1   Internal sources

Calcium sulfate is an important component of all Portland cements. Various forms of *calcium sulfate* (anhydrite, hemihydrate, dihydrate; sulfate-bearing industrial by-products) are added to clinker during cement grinding to enable control of cement setting characteristics. Sulfates are also known to accelerate hydration of the calcium silicates present in Portland clinkers, thus potentially increasing early strength of cement. Sulfate may be added in the form of natural or industrial-grade calcium sulfate dihydrate (gypsum) or anhydrite.

Additional sulfates originate from *clinker*, in which they are formed during clinker manufacturing from raw materials and fuel combustion products. They are predominantly present in clinker in the form of alkali- and calcium-alkali sulfates and, occasionally, in the form of calcium sulfate anhydrite or other phases (e.g. Skalny and Klemm 1981). The most common sulfate phases present in clinkers are arcanite, $K_2SO_4$, calcium langbeinite, $KC_2\bar{S}_3$, and aphthitelite $K_3N\bar{S}_4$. Important optical and crystal data for sulfate phases present in clinkers are well summarized in Tables 2.4 and 2.5 of Taylor (1997).

Please note that, with technological advances, the chemical and compound compositions of clinkers and cements, as well as other characteristics, may change. For example, during the last decades, the Blaine fineness of typical cements was increased dramatically, primarily in an attempt to increase concrete early strength. Increased fineness leads to higher early rate and heat of hydration, this fact potentially causing durability problems to be discussed later. In contrast, the amount of potential $C_3A$, as calculated by Bogue technique, remained for all practical purposes unchanged during the last fifty years or so. The reported changes in sulfate content and surface area may have an effect on the behavior of modern concrete exposed to chemical deterioration, including sulfate attack. Table 4.1 summarizes the changes in surface area and sulfate content of the five ASTM types of cements that occurred between 1950 and 1990.

The above changes in sulfate content are to some degree the result of environmental restrictions on sulfur and other emissions, resulting in changed burning and dust recycling practices. Thus, the proportion of sulfate that is present in typical cement in the form of alkali- and alkali-calcium sulfates, originating from the clinker, may be in some cases larger then it was in the past. Please note that, although the mean $SO_3$ content of the 1990 Type V cement was reported to be 2.3%, the range of $SO_3$ was reported to be 1.8–3.6%.

*Table 4.1* Comparison of mean fineness and sulfate contents of ASTM cements produced in years 1950 and 1990.

| ASTM Type | 1950 Fineness | 1990 Fineness | 1950 SO$_3$ | 1990 SO$_3$ |
|---|---|---|---|---|
| I | 332 | 369 | 1.9 | 3.0 |
| II | 338 | 377 | 1.7 | 2.7 |
| III | 477 | 548 | 2.3 | 3.5 |
| IV | 317 | 340 | 1.7 | 2.2 |
| V | 327 | 373 | 1.6 | 2.3 |

*Source*: Portland Cement: Past and Present Characteristics, *Concrete Technology Today*, PCA, July 1996

The overall increase in sulfate content and the changed mineralogy of the sulfates present in modern cements led to allegations regarding the unsuitability of certain high-sulfate cement concretes for applications utilizing heat-treatment (e.g. Hime 1996a,b; Collepardi 1997). Recent reevaluation of these charges does not confirm the above allegation (e.g. Klemm and Miller 1997; Skalny *et al.* 1997; Thomas 1998).

Occasionally, the source of sulfate in concrete may be the sulfates (e.g. gypsum) or sulfides present in *aggregate*. Typical example of a sulfide is iron sulfide (pyrite). In the presence of oxygen and moisture, pyrite may oxidize to form a low-pH sulfate solution that may contribute to concrete degradation by both *sulfate* and *acid attack*:

$$2FeS_2 + 7.5O_2 + H_2O \rightarrow Fe_2(SO_4)_3 + H_2SO_4$$

Sulfates may be components of mineral and chemical *admixtures*. When using an unknown or new admixture in concrete applications potentially exposed to sulfate conditions, it is advisable to check the chemical or mineralogical nature of these concrete components.

Finally, a source of internal sulfate may be the *mixing water*, but this is considered to be an improbable source of serious damage. However, please note that tap water in many localities has sulfate contents above 150 ppm, or even more. Such high concentration, especially in situations when such water is used for watering or otherwise in contact with concrete (external source of sulfate), could qualify the concrete for w/cm equal or below 0.50 (!).

### 4.2.2 External sources

The primary sources of external sulfate are natural sulfates of calcium, magnesium, sodium, and potassium present in *soils* or dissolved in *ground water*. Occasionally, sulfides may be present, such as pyrite, which may oxidize to form water-soluble sulfates. In arid areas, concrete structures in contact with ground water, having one face exposed to air, are particularly endangered due to increasing concentrations of sulfates at the air-exposed surface through evaporation of surface moisture. The probability of such increase in sulfate concentration decreases with concrete impermeability, i.e., decreasing water–cement ratio.

Additional sources of sulfate are solid *industrial wastes* exposed to precipitation or ground water. Examples are wastes from mining industry (e.g. cinders, oil-shale residues), combustion of coal (refuse from incineration, bottom ash, pulverized fuel ash), and metallurgical industry (e.g. low-quality blast-furnace slag). Ready availability of these sulfates to cause damage to concrete depends on their concentration and solubility, access and mobility of contact water, and environmental conditions.

Industrial or agricultural *waste waters* may contain sulfates too; for example, water from some cooling towers may under certain conditions reach relatively high sulfate concentrations. Agricultural waste is a well-known source of aggressive chemicals, although sulfate (e.g. from fertilizer) is not one of the most common ions present.

Finally, sulfates may originate from atmospheric pollution and, depending on the atmospheric conditions (temperature, humidity, wind, etc.), may contribute to increased sulfate concentration of the soil and ground water, thus increasing the potential for causing sulfate attack.

## 4.3    MECHANISMS OF SULFATE-RELATED DETERIORATION

In the following pages, an attempt is made to review in detail the importance of chemical mechanisms in sulfate attack. As mentioned above, these mechanisms are complex and in some cases not completely understood. Experimental data, mostly obtained on laboratory specimens, are generally in agreement with each other, but there is a variety of reported experimental data that are difficult to interpret and this led to some controversies. An attempt is made to discuss these disagreements and to suggest some explanations.

All sulfate attack mechanisms to be discussed are treated as physico-chemical processes that lead to certain physico-mechanical consequences. Such consequences include changes in porosity and permeability, volumetric stability, compressive and flexural strengths, modulus of elasticity, hardness, etc. – all these changes ultimately resulting in loss of durability and shortening of service life. Because expansive processes, specifically their occurrence and kinetics, are to some extent related to the conditions under which the expansion may occur, the needed relationships is discussed. Also discussed is the effects of environmental conditions and processing practices. Because of the existing uncertainties and controversies, the issues of so-called *delayed* ettringite formation (a form of *internal* sulfate attack) and so-called *physical sulfate attack* is also highlighted briefly under separate headings.

### 4.3.1    Sulfate attack mechanisms – a brief history

External sulfate attack in its classical form is known for many decades. As a matter of fact, Vicat, Le Chatelier, Michaelis and other European scientists conducted studies on sulfate resistance of cement-based concrete already in the nineteenth century. In North America, damage to concrete by external sulfates has been reported by American and Canadian engineers as early as in 1908. For example, Wig and coworkers (1915, 1917) reported on deterioration of concrete drain tiles exposed to alkali salts. They concluded that the observed deterioration was the consequence of inadequate processing (high porosity, low cement content) and resulted both from crystallization of salts in the concrete pores and from chemical reactions between cement constituents and the alkali salts from ground water. Serious studies on the effect of different clinker compositions on sulfate durability of concrete were initiated by the US Bureau of Reclamation, US Army Corps of Engineers, Portland Cement Association (PCA), the National Bureau of Standards (NBS, now NIST), the Canadian National Research Council (NRC), University of Saskatchewan and others as early as in the 1920s (DePuy 1997). It was soon recognized that the severity of the attack depends on the chemical form of the sulfate, magnesium sulfate being potentially more destructive than sodium sulfate, and both being more aggressive than calcium sulfate.

Field experience and laboratory studies have also shown that the sulfate resistance of concrete to expansive ettringite damage is improved by limiting the water–cement ratio and by lowering the $C_3A$ content of the clinker. This realization resulted in research to develop a sulfate resisting clinker, primarily by replacing the $C_3A$ clinker phase by the alumina-poor, ferrite-rich $C_4AF$. Early sulfate-resisting clinkers (Erzzement) were developed in Germany on the basis of early work by Michaelis (1901) and in Italy (Ferrari cement) on the basis of data by Ferrari (1919). The sulfate-resisting cement known today as ASTM Type V cement was first proposed and then adopted in the mid-1930s; some initial ideas on prevention of sulfate attack by cement composition can be traced back to some of the early works of Thorvaldson and coworkers (e.g. Swenson 1968; Thorvaldson 1952; Thorvaldson *et al.* 1925, 1927).

In the late 1920s, NBS and PCA funded research that resulted in the "Bogue" method of calculation of the main components of Portland cement. This method turned out to become an important tool for characterization of different cement compositions with respect to their durability in sulfate-rich environments. Although the Bogue method of calculation does not give absolutely correct mineralogical compositions, it remains a useful tool in enabling comparison of various clinkers.

The chemico-mechanistic aspects of sulfate attack were studied from the very early years of cement technology. The early researchers included Le Chatelier, Michaelis, Feret and others in Europe, and Bogue, Thorvaldson, Bates and many others in the United States and Canada.

## 4.4   TYPES OF EXPANSIVE REACTIONS

The interaction between sulfates on one hand and the constituents of unhydrated cement or hydrated cement paste on the other hand is very complex and depends on a variety of factors (Gartner and Gaidis 1989; Odler and Jawed 1991). A multitude of chemical reactions, taking place concurrently or subsequently, may be involved. The factors affecting the process include:

- the composition of the cement or hardened cement paste;
- the form of sulfate participating in the process;
- the type of interaction, i.e. whether the sulfate is a constituent of the cement or concrete mix undergoing hydration (internal sulfate attack)

or enters into the hardened paste/mortar/concrete from outside (external sulfate attack);
- temperature at which the interaction takes place, etc.

The reactions involved may result

- in an expansion of the paste;
- in chemical degradation of the present hydrate phases associated with a worsening of the intrinsic strength properties of the hardened cement paste;
- in crack formation within the hardened material, in surface scaling, spalling or delamination of the hardened material.

There is no doubt that an expansion of a cement paste occurs if any of the chemical reactions taking place within the paste is associated with an overall increase of the volume. So, a reaction schematically described by the equation:

$$A + B \rightarrow C$$

may cause expansion if:

$$V_A + V_B < V_C$$

where $V_A$ = volume of A
$V_B$ = volume of B
$V_C$ = volume of C

In cementitious systems, however, virtually all of the reactions in which sulfate ions are involved are associated with a *chemical shrinkage*, rather than with an expansion, which means that:

$$V_A + V_B > V_C$$

Some of the reactions of this type occuring in cementitious systems do not cause significant changes of the external volume, however other may or may not, depending on the conditions at which this reaction takes place.

Several theories have been forwarded to explain why a chemical reaction associated with a chemical shrinkage causes an expansion of a cement paste, which are discussed subsequently.

### 4.4.1   Increase of the solid volume

This theory assumes that an expansion of the cement paste may be caused by an increase of the volume of the present solids, even if the overall volume of the species involved becomes reduced. Such situation is typical in hydration reactions, which can be schematically expressed by the equation:

$$A_S \ + \ H_L \ \rightarrow \ AH_S$$

original     water     hydrated solid
solid

where S = solid

    L = liquid

$$V_A + V_H > A_{AH}$$

$$V_A < V_{AH}$$

where $V_H$    = volume of H

    $V_{AH}$ = volume of AH

Obviously, an increase of the solid volume will not cause an expansion in a fresh cement paste with sufficient plasticity that, instead, will reduce its external volume in line with the chemical shrinkage associated with the reaction.

In a hardened paste, under conditions of a random nucleation from the pore liquid, the nuclei of the newly formed solid phase will grow freely and will not exert pressure on the surrounding, as long as free space for crystal growth is available. A pressure on the surrounding walls may be only exerted if the crystals become too large and/or too numerous to fit into the available space. In cement pastes with "normal" water–cement ratios, however, the available pore space is large enough to allow a complete hydration of the binder, while still leaving behind a sufficiently large non-filled pore space. This explains why the hydration process that takes place in such pastes does not cause an increase of the external volume.

A distinct expansion of the paste would be conceivable only at extremely low water–cement ratios, as here the available pore space would not be large enough to accommodate the whole amount of hydrates formed in a complete hydration of the binder. However, rather than continue and cause an expansion, the hydration process tends to stop under these conditions before a complete hydration is attained. This is due to the mechanical restrains caused by the cohesion of the material that would need to be surmounted if the hydration process would continue.

### 4.4.2 Expansion in a topochemical reaction

*Topochemical reactions* are very broadly defined as chemical reactions in which the solid phase produced in the process is formed in the space originally occupied by one of the starting solid materials. Solid phase transitions, such as the conversion of $\beta$-dicalcium silicate to $\gamma$-dicalcium silicate clearly meet this definition. Under this broad definition many hydration reactions can also be considered topochemical, as for example reactions in which water molecules enter the crystalline lattice of the original material. An example of such reaction is the conversion of the calcium aluminate hydrate $C_4AH_{12}$ to $C_4AH_{19}$, in which an additional layer of water molecules is inserted in each unit cell within the crystalline lattice.

According to a stricter definition of topochemical reactions, the product of reaction must form a true crystalline overgrowth on the reactant with matching d-spaces, in addition to be formed in the space occupied by the reactant. However, such a strict definition is being only rarely used in cement chemistry.

In cement chemistry a hydration reaction may be called *topochemical* even if the starting material becomes dissolved first, as long as the reaction product precipitates only within the space that became available by the dissolution process or in its immediate vicinity.

In contrast to a topochemical reaction, in a process called commonly a *through-solution reaction*, the reaction product precipitates randomly from the liquid phase, after sufficient amounts of the starting material became dissolved and a sufficiently high degree of oversaturation was attained. Also, a precipitation of a solid reaction product from the liquid phase onto the surface of another solid phase does not constitute a topochemical process.

Topochemical hydration reactions are favored if the capability of the reaction product to crystallize is markedly greater than the rate of dissolution of the starting anhydrous material. Under such conditions the dissolved ions do not have time to migrate away into the liquid phase but, instead, the reaction product precipitates immediately at or close to the surface of the anhydrous phase. As the reaction progresses, the topochemical product formed in this way tends to partially block access to the underlying anhydrous phase, but, nevertheless, the surrounding liquid phase continues to migrate through the pores of the already formed hydrate layer towards the residual starting material. Thus additional amounts of the hydrate phase are formed at the anhydrite–hydrate interface, usually at a parabolically decreasing rate, until the starting material becomes completely replaced by the reaction product.

Due to its larger mass and lower density the hydrate formed in situ occupies a substantially greater volume than the "parent" anhydrous solid phase and thus necessarily exceeds the original boundaries of the latter. Thus, the newly formed solid can give rise to a local expansion. If this expansion occurs in the restricted space of a porous solid, such as in a hardened cement paste, local stresses distributed randomly within the hardened material may be generated. These in turn may cause an increase of the external volume of the material, or even formation of cracks. An example of a topochemical hydration is the conversion of periclase [MgO] to brucite [$Mg(OH)_2$] taking place in the hydration of cements with a high MgO content.

Please note, that the same chemical reaction may or may not proceed topochemically, depending on the reactivity of the starting material. For example, the hydration of lime [CaO] to portlandite [$Ca(OH)_2$] proceeds topochemically if the starting CaO had been produced by a decomposition of $CaCO_3$ at a high temperature, and proceeds in a through-solution reaction if the decomposition temperature was lower. This different behavior is due to differences in the dissolution rate of the two CaO materials, brought about by the different burning temperatures employed. CaO that had been formed at lower temperatures is namely microcrystalline, possesses a higher specific surface area and thus dissolves faster than lime burnt at higher temperatures that is distinctly coarser and has a more orderly arranged crystal lattice (e.g. Odler 2001).

It must also be mentioned that a topochemical reaction may cause an expansion independently on the degree of crystallinity of the reaction product and without any uptake of water from the environment, even though all hydration reactions of which we are aware – as already mentioned – are associated with an overall chemical shrinkage.

### 4.4.3   Oriented crystal growth

An oriented anisotropic crystal growth may develop if a crystalline reaction product is formed in a topochemical reaction at the interface between the anhydrous parent phase and the hydrate. Under these conditions the newly formed amounts of the hydrate phase exert pressure on those already in existence and formed earlier. This may cause a situation in which the crystals of the reaction product grow and may exert pressure in a direction perpendicular to the original solid–liquid interface. Such form of crystal growth is associated with an increase of porosity and thus with an increase of the external volume, or even crack formation within the set cement paste. The expansion may be particularly significant if the reaction product possesses an acicular morphology and grows in its longitudinal direction. Obviously, an expansion of the cement paste gets under way only when the expanding crystals come into contact with, and exert pressure on each other. An external expansion may take place without an uptake of water from the environment.

### 4.4.4   Expansion caused by crystallization pressure

Crystals formed in a crystallization from a supersaturated pore solution in a porous body may generate pressure on the pore walls after reaching an appropriate size (Chatterji and Thaulow 1997; Ping and Beaudoin 1992; Scherer 1999 and others). The maximum pressure exerted by a single crystal may be expressed by the equation (Brown and Taylor 1999; Scherer 1999):

$$p = RT/V \cdot \ln C/C_0$$

$p$ = pressure in MPa
$R$ = gas constant $8.3\ \mathrm{J} \cdot \mathrm{K}^{-1} \cdot \mathrm{mol}^{-1}$
$T$ = temperature in °K
$V$ = molar volume in $\mathrm{m}^3 \cdot \mathrm{mol}^{-1}$
$C$ = concentration of the solute
$C_0$ = equilibrium solubility

From the equation it is obvious that the degree of over-saturation of the solute is the most important factor determining the magnitude of the pressure exerted on the pore wall, however other factors are also involved (Ping and Beaudoin 1992; Scherer 1999 and others).

The pressure that may be exerted by a crystal growing into a pore depends also on the contact angle between the crystal and the pore wall, determined by the curvature of the crystal–liquid interface. If the contact angle is small, the stress will be reduced, whereas the maximum pressure will be exhibited if the crystal is completely *non-wetting*.

The crystallization pressure depends also on the size of the pore and becomes smaller as the size of the pore increases.

If the growing crystal is faced with an incomplete restrain, (such as by a pore wall) it preferentially grows on its unrestrained face (i.e. towards the liquid phase). Thus, an external expansion may only take place if the solid product forms and grows in a confined space.

When an acicular crystal grows across a pore, stresses may be exerted at each of its ends. Two factors, however, may interfere: The crystal may yield under the generated stress: If this happen, the crystal will grow laterally in the pore, rather than across its cross-section. Alternatively, the crystal may buckle under the load generated by its own growth.

Under real condition an expansion and noticeable deterioration of a cement paste cannot be generated by the crystallization of a single crystal in a single pore, because the generated stress, even if large, acts on a too small volume. To produce expansion and ultimately cracking, crystal growth must take place in a large enough volume of the pore space, so that also the largest flaws present in the paste, that ultimately determine the cohesion of the material, are included.

### 4.4.5  Expansion caused by swelling phenomena

Systems possessing a hydrophylic surface may take up water from the environment by surface adsorption, resulting in an increase of the external volume and generation of an expansive pressure. This effect depends greatly on the specific surface area of the material and becomes significant only at high specific surface areas.

Other systems may take up water by incorporating additional water molecules into their crystalline lattice, most often into interlayer spaces. Also this phenomenon is usually accompanied by an increase of the external volume, that may be in some instances significant.

### 4.4.6  Other expansive processes

Thorwaldson (1952) presented a theory according to which an expansion within a cement paste may be caused by *osmotic phenomena*. In this theory it is assumed that the hydration product forms a semi-permeable membrane around the grains of the starting anhydrous material which hinders the free migration of ions between the liquid in immediate contact with the surface and the bulk solution. The resultant difference in the concentration of the two sides of the membrane generates difference in osmotic pressure, and this in turn may generate internal stresses and ultimately an increase of the external volume of the paste.

The possibility of an expansion caused by a *reversal of local desiccation* has also been discussed (Brown and Taylor 1999). On disputed mechanism assumes the initial formation of the ettringite phase in a partially dehydrated state, due to a limited availability of water. An expansion occurs when subsequently this phase takes up additional water from an external source. Alternatively it is assumed that, to produce fully hydrated ettringite, water is taken up from the surrounding C-S-H and an expansion occurs when the latter retakes water from the environment.

## 4.5 ETTRINGITE FORMATION AND EXPANSION

Ettringite, a crystalline AFt compound of the formula $[Ca_3Al(OH)_6 \cdot 12H_2O]_2 \cdot (SO_3)_3 \cdot 2H_2O$ or abbreviated $C_3A \cdot 3C\bar{S} \cdot 32H$, plays a crucial role in cement chemistry and may affect the properties of a hardened cement paste positively or negatively. The role of this phase in the hydration of Portland cement is discussed in Chapter 2.

Ettringite may precipitate from a water solution that contains the ions $Ca^{2+}$, $Al(OH)_4^-$, $SO_4^{2-}$ and $OH^-$ in concentrations making the liquid supersaturated with respect to this phase. The solubility product of ettringite at $25\,°C$ was found to be as follows (Damidot and Glasser 1992):

$$K_{sp} = (Ca^{2+})^6 \cdot (Al(OH^-)_4)^2 \cdot (SO_4^{2-})^3 \cdot (OH^-)^4 = 2.80E-45$$

Invariant points in the system $Ca-Al_2O_3-SO_3-H_2O$ related to ettringite are given in Table 4.2.

Pure ettringite is stable up to about $90\,°C$ (Taylor 1997), however, in Portland cement pastes, it is not typically formed at temperatures above about

*Table 4.2* Invariant points in the system $CaO-Al_2O_3-SO_3-H_2O$ related to ettringite.

| Solid phases | Concentration (mmol/l) | | |
|---|---|---|---|
| | CaO | $Al_2O_3$ | $CaSO_4$ |
| Ettringite, gypsum, $Ca(OH)_2$ | 20.7 | 0.0272 | 12.2 |
| Ettringite, gypsum, hydrous alumina | 0.151 | 0.103 | 15.1 |
| Ettringite, monosulfate, $Ca(OH)_2$ | 21.1 | 0.022 | 0.0294 |
| Ettringite, monosulfte, Hydrous alumina | 5.98 | 1.55 | 5.88 |

*Source*: Taylor (1977)

60–70 °C. Ettringite in suspensions is stable only above about pH = 10–11. The presence of KOH in the liquid phase in concentrations above 0.5 mol may prevent its precipitation (Brown and Bothe 1993), which most likely is due to the reduction of the activity of water in such concentrated solutions. Thermodynamic data related to ettringite were published by Brown (1986) Brown and Bothe (1993), Clark and Brown (1999), Damidot and Glasser (1992, 1993) and others.

In addition to a precipitation from the liquid phase, ettringite may be also formed in water suspensions or pastes from a variety of compounds of aluminum, in combination with calcium sulfate, in the presence of calcium hydroxide. All these reactions are associated with a chemical shrinkage of different magnitude.

In the hydration of Portland cement and related calcium silicate based binders ettringite is formed from tricalcium aluminate acting as source of aluminum oxide:

$$C_3A + 3C\bar{S}H_2 + 26H \rightarrow C_6A\bar{S}_3H_{32}$$

The reaction is also relevant in internal sulfate attack. The chemical shrinkage associated with the reaction is 73.1 ml/mol $C_3A$ or 9.37%. This reaction may cause an expansion of a Portland cement paste, but only if substantial amounts of ettringite are formed after the paste had already set and gained a measurable strength.

In external sulfate attack the expansion reaction most important is the formation of ettringite from monosulfate:

$$C_4A\bar{S}H_{12} + 2C\bar{S}H_2 + 16H \rightarrow C_6A\bar{S}_3H_{32}$$

Here the chemical shrinkage is 39.3 ml/mol $C_4ASH_{12}$ or $\bar{5}.27\%$.

In some expansive cements and in Ca-sulfoaluminate based binders ettringite is formed from tetracalcium trialuminate sulfate:

$$C_4A_3\bar{S} + 8C\bar{S}H_2 + 6CH + 74H \rightarrow 3C_6A\bar{S}_3H_{32}$$

The chemical shrinkage associated with this reaction is 40.5 ml/mol $C_4A_3\bar{S}$ or 1.87%. This reaction may or may not cause an expansion, depending on the degree of mechanical restrain and possibly also on the $SO_3/Al_2O_3$ ratio in the cement.

If allowed to hydrate in paste form in combination with pure $C_3S$ in amounts sufficient to produce 30% ettringite, mixes made with $C_3A$, $C_4A_3\bar{S}H_{12}$, $C_4A_3\bar{S}$ and also $C_4AF$ and CA exhibited a distinct expansion; in contrast mixes made with aluminum sulfate and a $CaO–Al_2O_3–SiO_2$ glass did not (Odler and Collan-Subauste 1999). The mix samples that expanded did so at different rates, the expansions of those made with $C_4A\bar{S}H_{12}$ (AFm), $C_4A_3\bar{S}$ and CA being particularly fast. The extent of expansion depended also on the availability of water from the environment, increasing in the order: samples cured in dry air < samples cured in water vapor saturated air < samples cured under water. From Figure 4.1, in which the attained expansion is plotted as function of the amount of ettringite formed, it may be seen that the capacity of the individual starting Al-phases to produce expansion varied greatly. From the figure it is also apparent that a marked expansion got under way only after the formation of a threshold amount of ettringite.

As in all expansive hydration processes the relevant chemical reaction is paradoxically associated with a chemical shrinkage (caused by the fact that the decrease of the volume of the liquid phase is greater than the increase of the volume of the solid phases) the expansion cannot be explained by a (non-existing) increase of the overall volume of the phases involved. Also, the expansion cannot be interpreted just by the increase of the solid volume accompanying the process, for reasons discussed above. It may be mentioned in this connection that in the hydration of other phases commonly present in cements (such as in the hydration of $C_3S$ yielding C-S-H and $Ca(OH)_2$), the accompanying increase of the solid volume does not results in an external expansion.

Scanning electron microscopy observations of expansive cementitious systems often reveal the presence of acicular ettringite crystals growing perpendicularly from the surface of the anhydrous Al-compound from which it is being formed. This fact seems to support the "oriented crystal growth"

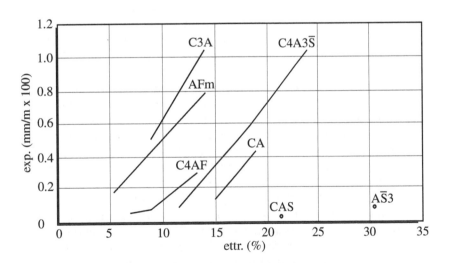

*Figure 4.1* Expansion of specimens made from an aluminate, $C_3S$ and gypsum as a function of the amount of formed ettringite.

Source: Reprinted from *Cement and Concrete Research*, vol. 29, Odler, I. and Colan-Subauste, J. "Investigation on cement expansion associated with ettringite formation", pp. 731–735, Copyright 1999, with permission from Elsevier Science

theory (Bentur and Ish-Shalom 1974–1975; Herrick *et al.* 1992; Moldovan and Butucescu 1980; Ogawa and Roy 1981–1982; Wang *et al.* 1985).

Presently, most but not all investigators consider the topochemical ettringite formation around a precursor phase and an oriented growth of this reaction product to be the main, or even the sole reason for a sulfate related expansion in calcium and and aluminate-containing cementitious systems. (Alunno-Rosetti *et al.* 1982; Bentur and Ish-Shalom 1975; Brown and Taylor 1999; Clastres *et al.* 1984; Deng and Tang 1994; Kalousek and Benton 1970; Lafuma 1927; Mather 1973; Odler and Gasser 1988; Ogawa and Roy 1980–1981; Taylor 1997; Wang *et al.* 1985 and others). Such mechanism does not presume necessarily the presence of ettringite crystals visible by SEM in the paste and, as will be elaborate later, finely dispersed ettringite is that form of this phase that is commonly formed in sulfate expansion.

To bring about a topochemical formation of ettringite and thus produce expansion, it is also generally agreed that sufficiently high concentrations of $Ca(OH)_2$ must be present in the liquid phase. (Alunno-Rosetti *et al.* 1982; Clastres *et al.* 1984; Degenkolb and Knoefel 1996; Deng and Tang 1994; Mehta 1973a). At high calcium hydroxide concentrations, namely, aluminate ions cannot migrate too far from the phase serving as aluminum source, due to supersaturation of the liquid phase relative to ettringite (and in the absence of sulfate ions also relative to calcium aluminate hydrate phases). Thus ettringite will be formed preferentially at the surface of the Al-source in a topochemical reaction. If, on the other hand, the $Ca(OH)_2$ concentration in the liquid phase is low, $Al(OH)^{4-}$ ions can migrate more freely, and ettringite can precipitate randomly from the liquid phase without generating expansive stresses. It was estimated that ettringite tends to be formed in a through-solution reaction at $Ca(OH)_2$ concentrations below 0.020 mol/l and topochemically, if this concentration is higher (Deng and Tang 1994).

In addition to the mode of hydration, but probably for the same reason, the presence of calcium hydroxide in the liquid phase affects also the morphology and size of the ettringite crystals formed. While well developed ettringite crystals, up to 6 μm long, are formed in the absence of calcium hydroxide or at its very low concentrations, small, almost colloidal particles, only about 1 μm large, form if the liquid phase is saturated with respect to $Ca(OH)_2$ (Mehta 1973a,b). Thus, in cement pastes undergoing hydration, in which the $Ca(OH)_2$ concentration is high, ettringite is formed at the surface of the tricalcium silicate phase in form of a fine, nearly colloidal layer. Also, in mature pastes that had expanded as a consequence of a reaction between the AFm phase, produced before, and additional sulfate ions, the ettringite formed in this reaction is present in form of crystals of or below-micrometer dimension, intimately mixed with the C-S-H phase (Brown and Taylor 1999).

It was suggested, that the amount of alkalies in proportion to the amount of $C_3A$ and sulfate ions, may also affect the extent of expansion taking place in cement pastes. According to Aardt and Visser (1985) an expansion does not take place or is significantly reduced if the individual oxides are present in the following ratios:

$$C_3A/(SO_3 + Na_2O_{eq}) < 3 \quad \text{and} \quad SO_3/Na_2O_{eq} > 1 \text{ but} < 3.5$$

In recent years a controversy has arisen concerning the role of relatively large crystals of ettringite that are often found in gaps at the aggregate-cement paste interface in pastes that had been subjected to heat curing. According to one view, these gaps are formed due to the crystallization pressure generated by the growth of the ettringite crystals themselves; an opposite view maintains that ettringite is only re-precipitated into already existing spaces. This subject is discussed in Section 4.3.7.1 that deals with delayed ettringite formation.

Under field conditions involving repeated wetting and drying ettringite may precipitate in voids resulting from accidentally entrapped air, or produced by air entraining admixtures that were added to the concrete mix to improve its frost resistance. Such precipitation by itself appears not to be deleterious, as the crystallization pressure exerted across the air void is probably too low, to cause an expansion of the hardened paste. However, there are conflicting views regarding the effectiveness of fully or partially filled air voids as to their ability to protect concrete from damage caused by exposure to freezing temperatures (Detwiler and Powers-Couche 1996; Hulshizer 1997; Ouyang and Lane 1997 and others).

According to a hypothesis forwarded by Mehta and coworkers (1973a,b; 1978, 1982) the sulfate-related expansion of concrete is caused by water uptake and swelling of "colloidal" or, more exactly said, "micro-crystalline" ettringite that may be already present in the cement paste, rather than by the actual formation of this phase. The basis for their hypothesis was the observation of a distinct expansion of compacts made from "colloidal" ettringite and $C_3A$-free Portland cement (Mehta and Wang 1982). However, from their measurements of expansion on ettringite compacts Alunno-Rossetti and coworkers (1982) concluded that the swelling of ettringite alone could not account for all the volume increase associated with sulfate expansion, and that a simultaneous development of crystallization pressure must have also been involved.

An expansion resulting from swelling of ettringite would involve an uptake of water by this phase. Experiments performed both on Portland cements (Mather 1973; Odler and Gasser 1988) and $C_3S$-based pastes (Odler and Collan-Subauste 1999) with high sulfate contents revealed that expansion may take place even when no water is taken up by the sample from the environment. At the same time it was observed, however, that the extent of expansion increases if the material is kept in humid air or under water, rather than in dry air.

The swelling hypothesis of Mehta and coworkers has been recently discussed by Brown and Taylor (1999). They pointed to the fact that an imbibition of water and a resultant swelling involves gel-like materials of indefinite composition which have the flexibility to expand without breaking, and ettringite does not appear to exhibit these necessary features. Ettringite not previously dried cannot take up water internally as there is no room for more water molecules in its crystal structure. At the same time it cannot bind significant amounts of water on its surface either, as its specific surface area remains low, even if present in form of relatively small crystals. Paradoxically, The C-S-H phase, which has a significantly higher surface area than micro-crystalline ettringite and has the capability to bind significant amounts of water, does not swell.

The specific mechanism of additional expansion associated with ettringite formation under conditions of water uptake from the environment is not obvious. However, it appears unlikely that this phenomenon can be attributed to an increase in the volume of the present ettringite crystals themselves.

In conclusion, it may be said that the formation of ettringite in cement pastes may or may not be associated with an external expansion of the material. Several conditions must be met, to produce expansion, including the following:

- The amount of ettringite produced must exceed a threshold amount, so that a pressure may be generated on the neighboring solids.
- Ettringite must be formed in a topochemical process, resulting in an oriented growth of the ettringite crystals towards the neighborly solids. (The question of whether and under which conditions the crystallization pressure that is generated when ettringite crystallizes from a supersaturated pore solution suffices to produce an expansion, is a subject of discussion)
- Only ettringite that had been formed after the cement paste, mortar or concrete had set, has the capacity to generate expansion. Because of this, aluminum sources that react with calcium sulfate to form ettringite too fast, such as Al-sulfate, may not cause expansion.

Please note that in concrete mixes in which the strength exceeds an optimum range and the material lost most of its deformability, crack formation, rather an expansion may become the predominant form of damage associated with ettringite formation.

## 4.6   OTHER FORMS OF SULFATE-RELATED EXPANSION

*Gypsum* (calcium sulfate dihydrate) may be formed in external sulfate attack in a reaction between the sulfate ions entering the cement paste in form of a water-soluble sulfate compound and calcium hydroxide of the cement paste:

$$2Me^+ + SO_4^{2-} + Ca^{2+} + 2OH^- \rightarrow CaSO_4 \cdot 2H_2O + 2Me^+ + 2OH^-$$

The question, whether or not such a reaction may or may not cause an expansion of the hardened cement paste has been extensively discussed and the opinions on this subject have been contradictory (Bing and Cohen 2000; Dimic and Drolc 1986; Hansen 1963; Mather 1996; Mehta 1992; Ping and Beaudoin 1992; Wang 1994; Yang *et al.* 1996 and others). Recently Tian and Cohen (2000) demonstrated in experiments performed with pure tricalcium silicate pastes that a formation of gypsum may cause a significant expansion even when a formation of ettringite can be excluded. The mechanism by which the formation of gypsum generates expansion is not obvious. It has been suggested that under appropriate conditions, such as at a high degree of supersaturation of the liquid phase with respect to gypsum, the generated crystallization pressure may be high enough to generate expansion of the paste (Chatterjee 1968; Ping and Beaudoin 1992 and others). It is also conceivable, however, that the expansion is due to a topochemical reaction between sulfate ions present in the pore liquid and crystalline calcium hydroxide constituting the cement paste.

In some concrete structures exposed to sulfate solutions *thaumasite* [$CaSO_4 \cdot CaSiO_3 \cdot CaCO_3 \cdot 15H_2O$] was found as the main sulfate-bearing phase formed. The role of this phase in external sulfate attack will be discussed in Section 4.9.6.

If attacked with sodium sulfate in high concentrations the so-called *U-phase* may be formed and its formation may also cause a degradation of the cementing material. This subject will be discussed in Section 4.9.1.

In concretes that were made with an aggregate that contains *pyrite* [$FeS_2$] this constituent may undergo an oxidation with oxygen of air, converting to iron hydroxide [$Fe(OH)_3$] and sulfate ions (Casanova *et al.* 1996). This process is associated with an increase of the solid volume and may generate an expansion of the hardened concrete. The produced sulfate ions react with calcium hydroxide, formed in the hydration of the cement, to yield gypsum. The latter phase may react subsequently with any present tricalcium aluminate or monosulfate to yield ettringite, resulting in a further expansion of the material.

Chatterji and Jeffery (1963, 1967) suggested that the expansion in an external sulfate attack is due to the formation of the monosulfate phase from calcium aluminate hydrate via an ion exchange process. The latter phase is formed in the paste from residual $C_3A$, after all the $SO_3$, originally present in cement, had been consumed.

Finally, it was suggested that expansive forces may develop in a cement paste as a result of a conversion of "amorphous" to "crystalline" calcium hydroxide, brought about by a catalytic action of sulfate ions (Chatterji 1968) An existence of such mechanism has not been experimentally confirmed.

## 4.7 INTERACTION OF SULFATES WITH THE C-S-H PHASE

In addition to a chemical interaction with the Al-bearing phases of Portland clinker and their hydration products, sulfate ions can also interact with the

C-S-H phase formed in the hydration of the dicalcium- and tricalcium silicate phases of cement. In laboratory experiments on gypsum + tricalcium silicate blends strength losses of up to 40% were observed (Bentur 1976; Mehta *et al.* 1979). In Portland cement the attained strength increases with increasing gypsum content to an optimum value and starts to decline when the optimum is exceeded. At too excessive sulfate contents the capability of the material to resist expansive forces caused by simultaneous ettringite formation may decline, due to loss of strength.

In external sulfate attack the migration of sulfate ions into the concrete body may be accompanied by a gradual dissolution of portlandite and decomposition of the C-S-H phase, constituting the hydrated cement paste. In the latter case the C/S ratio of this phase gradually declines as increasing amounts of $Ca^{2+}$ are removed from the structure. This process results in a gradual decline of the binding capability and thus in a strength loss of the hardened paste. Reactions of this type may take place if concrete is attacked by alkali sulfates, but are particularly severe if ammonium sulfate or even pure sulfuric acid act as corrosive agents. These subjects are discussed in Sections 4.9.4 and 4.9.5.

A special form of C-S-H degradation takes place, if concrete is attacked by a magnesium sulfate solution. The mechanism of this process will be discussed in Section 4.9.3.

## 4.8   INTERNAL SULFATE ATTACK

Most national and international standards put restrictive limits on both sulfate and $C_3A$ contents of cements, the purpose of this double restriction being, among others, prevention of *internal* sulfate attack. The importance of such restriction was recognized when it was observed that extensive expansion might occur in the absence of external source of sulfate. The recent ASTM standard for Type V sulfate resisting cement limits the $C_3A$ content to 5%, the $C_4AF + 2(C_3A)$ content to 25%, and the $SO_3$ content to 2.3% (ASTM 1995). Similar restrictions are imposed by other standards. Please remember that $C_4AF$ and $C_3A$ contents calculated by the Bogue-equations may not be in line with experimentally obtained values (e.g., obtained by quantitative X-ray diffraction or optical microscopy), thus the relationship between clinker composition and sulfate resistance is somewhat uncertain. Sulfates may also originate from aggregate or other concrete components, thus composition of all concrete ingredients, including mix water, should be considered in the mixture design.

Cases of internal sulfate attack in properly designed, cured and maintained concrete are uncommon. However, taking into account the relatively high mean clinker sulfate content of modern cements (Gebhardt 1995; PCA 1996), the potential for internal sulfate attack conditions should be closely monitored (see Table 4.1). This is especially important in situations when increased levels of high-sulfate cements are used in concrete processed by steam curing and placed at elevated temperatures, or both. Although some of the details of the "sulfate concentration versus processing temperature" relationship are unknown at the present time, observation of the heat-induced internal sulfate attack form, referred to in the literature as *delayed ettringite formation* (DEF), are serious enough to warrant more detailed discussion.

## 4.8.1 Internal sulfate attack at ambient temperature

In internal sulfate attack at ambient temperature the deterioration of hardened concrete is brought about by the action of sulfates present in the original mix in excessive amounts, or of those formed from sulfur compounds other than sulfates, also present in the starting material.

The main sulfate constituent in Portland cement is calcium sulfate, interground to the cement in the course of manufacturing. It is usually present in form of hemihydrate or anhydrite and rarely also as dihydrate. Distinct, but limited amounts of sulfate ions are also incorporated into the crystalline lattices of clinker minerals. Most of the sulfate ions are incorporated in the belite phase followed by alite and tricalcium aluminate, whereas the amount present in the ferrite phase is small enough to be ignored. For clinkers with up to 3% $SO_3$, the concentrations are below 0.5% in alite and rarely exceed 2.0% in belite (Taylor 1999). Sulfates bound within the clinker minerals are released into the liquid phase slowly and thus have the potential to act deleteriously, however only at excessively high overall $SO_3$ contents in the mix. Additional amounts of $SO_3$ may be bound within the clinker in form of well soluble and thus potentially less damaging sulfates, namely arcanite [$K_2SO_4$], aphthidalite [$3K_2SO_4 \cdot Na_2SO_4$], thenardite [$Na_2SO_4$] and calcium langbeinite [$K_2SO_4 \cdot 2CaSO_4$]. Some anhydrite [$CaSO_4$] may be also present. In some special cements, such as in sulfo-belite cement, the sulfate bearing phase $C_4A_3\bar{S}$ may also be present.

In composite (blended) cements, additional amounts of $SO_3$ may be present in the latent hydraulic or pozzolanic additive present in the binder in addition to the clinker and calcium sulfate. Some ashes that are produced in combustors designed to remove sulfur dioxide from the flue gases, may contain significant amount of sulfur, partially or completely in form of calcium sulfite rather than calcium sulfate.

The total amount of $SO_3$ that may be present in Portland cement, must not exceed an upper limit to prevent the hardened cement paste from unwanted expansion. In the US ASTM standard C 150-94, the highest acceptable overall $SO_3$ content for Portland cements varies between 2.3 and 4.5%, depending on the cement type.

In concrete mixes, additional amounts of $SO_3$ may be introduced as constituents of the aggregate employed. Obviously, rocks containing distinct amounts of $SO_3$ may be used as concrete aggregate without concern only if the sulfate is present in a sufficiently insoluble form such as, for example, in the form of baryte [$BaSO_4$], whereas aggregates containing even a moderately soluble sulfate phase, such as anhydrite [$CaSO_4$], must not be used. Also, some rocks may contain sulfur in sulfide form, such as pyrite [$FeS_2$]. If used as concrete aggregate, a gradual oxidation of this constituent to sulfate by the oxygen of air may cause an expansion of such aggregates, and act deleteriously even before a chemical interaction with constituents of the cement may get under way (Casanova *et al.* 1996).

Finally, $SO_3$ present in the mixing water may also contribute to internal sulfate attack, but a contribution if this constituent of the concrete mix is in most instances negligible and may become noticeable only at extremely high $SO_3$ concentrations.

In mixes made with cements whose $SO_3$ content lies within admitted limits, with an aggregate whose soluble $SO_3$ content is negligible and with "normal" water, no $SO_3$ related expansion and concrete damage takes place. Under these conditions, due to the exhaustion of the supply of available $SO_3$, the formation of ettringite is virtually completed at about the time of setting, i.e. prior to the stage at which the formation of this phase may generate any stresses within the material. On the other hand, if the amount of $SO_3$ in the mix is excessive, the formation of ettringite may continue even at later stages (after setting) and stresses develop which may cause expansion and even cracking of the concrete.

### 4.8.2   Heat-induced sulfate attack or DEF

The term *delayed ettringite formation* (DEF) became a *terminus technicus* in the late 1980s, when Heinz and Ludwig published a series of papers on expansion of laboratory and field mortars and concretes that were exposed to elevated temperatures and subsequently cured at room temperature under moist conditions (Heinz 1986; Heinz and Ludwig 1986, 1987). Since then, hundreds of papers were published on both laboratory studies performed to elucidate the mechanisms involved and on damaged concrete structures to explain the experimental findings in the field. For more information, see reviews by Day (1992), Lawrence (1995b), Thomas (1998), RILEM (2001), and Kjellsen *et al.* (1991), among others. DEF was alleged to be the culprit in several complex and expensive litigations in Canada and the United States, and several conferences were held to discuss the outstanding issues. For detailed information check professional literature (e.g. Johansen *et al.* 1993, 1995; Federal Supplement 1995; Mielenz *et al.* 1995; Scrivener 1996; Roberts and Hooton 1997; Erlin 1999).

As is often the case in discussion of newly described phenomena, the terminology and nomenclature used in the literature to describe the DEF-induced deterioration is confusing. *First* of all, it should be recognized that the phenomenon is clearly an *internal* form of sulfate attack because it is triggered by sulfate that is internal to concrete. All necessary ingredients (sulfate, alumina, calcium, and water) are present in the system at the time the concrete is produced, although additional water may be required to produce the damaging expansion; such water is supplied during post-curing storage or field use. Glasser (1996) refers to such processes as being "isochemical" because the bulk composition of the system remains unchanged. The process is *internal*, notwithstanding the fact that the actual expansion of the concrete matrix may theoretically be triggered by at least two mechanisms: either by excess of internal sulfate (cement, aggregate, admixtures, etc.) or by decomposition and subsequent reformation of all or part of the ettringite present in the system.

*Second*, the available literature uses numerous expressions to describe the phenomenon: secondary ettringite formation, late ettringite formation, delayed ettringite formation (DEF), reversionary ettringite formation, heat-induced internal sulfate attack, heat-induced mortar expansion (HIME), etc. For some, *delayed* and *secondary* means the same. *Secondary* ettringite, in addition of being mistakenly called *delayed*, is also called *harmless*. For others, *late* is a sub-category of *delayed*. *Primary* ettringite is also called *good* and *normal* ettringite but, at the same time, is recognized as the ettringite that may cause expansion. In the authors' view, the use of the expression "secondary ettringite formation" to describe the heat-induced effect is an incorrect choice, because it is now obvious that the actual expansion is not caused by formation of the "secondary" ettringite. In a similar way, the expression "delayed" ettringite formation does not make it clear whether the "delayed" action is caused by the presence and release of slowly-soluble sulfates from the clinker or by slow reformation of previously heat-decomposed primary ettringite. According to Mehta (2000), *all* ettringite-related expansion is caused by *delayed* reaction. The above problems with the nomenclature show that a serious attempt should be made to assign to the phenomenon a name that correctly describes the particular chemical mechanism and is not confusing with respect to the terminology used to describe other processes or phenomena. The issue of ettringite nomenclature has been discussed by Odler (1997).

*Third*, there is some confusion regarding the use of the terms *primary* and *secondary* in relation to ettringite or other phases formed in cementitious materials. We use the expression *primary* for a phase formed as the immediate product of a chemical reaction. In a cementitious material, *primary* products will normally, and perhaps always, be distributed as submicroscopic crystallites or gelatinous material within the cement paste. Examples would thus include the C-S-H, calcium hydroxide or ettringite formed as hydration products of the clinker phases, and ettringite crystals of or below micrometer dimensions formed through external sulfate attack or DEF. We shall use the expression *secondary* for phases formed by recrystallisation of primary phases, examples thus including calcium hydroxide or ettringite present in cracks or other cavities within the paste or elsewhere in a mortar or concrete. Because recrystallisation usually results from Ostwald ripening, crystalline secondary phases will normally be composed of larger crystals than the primary phases from which they have been formed. Such use of terms *primary* and *secondary* appears to be consistent with that commonly used in petrology.

The formation of primary phases can under some circumstances cause expansion. There have been differences of opinion as to whether expansion can result from formation of secondary phases. We consider this unlikely, as explained earlier.

The above examples show clearly that the phenomenon of ettringite expansion is complex and that what is needed is development of internationally acceptable nomenclature based on sound *mechanistic understanding* of the relevant processes. In our view, the two *internal sulfate attack* mechanisms – excess sulfate-generated expansion and heat treatment-generated expansion – should not be confused and should be assigned different names – notwithstanding the fact that the *expansion* mechanisms themselves may be the same or related. Our preference would be to talk about *composition-induced internal sulfate attack* and *heat-induced internal sulfate attack*, respectively. These and other related issues are being considered at the present time by the RILEM Technical Committee on Internal Sulfate Attack (TC-ISO).

Unless otherwise stated, the following mechanistic discussions will relate solely to heat-induced formation of ettringite, or DEF.

### Visual observations

Available information shows that heat-induced sulfate attack (DEF), in line with all localized sub-microscopic volume increases, is characterized by map cracking, longitudinal cracking, occasional warping of the concrete structure or product, and some spalling. Because in the majority of the know cases internal sulfate attack and ASR were operative simultaneously, it is difficult to make any definitive statement credibly characterizing the "DEF" damage caused by sulfate reactions alone.

Figure 1.3 shows a railroad tie allegedly damaged by ASR or heat-induced sulfate attack or both. Note the pattern or map cracking at the shoulders of the tie (a) and the longitudinal cracks between the rail shoulders (b). Similar pattern of cracking has been observed in large-size prefabricated beam-boxes and other structures (also subjected to simultaneous ASR and heat-induced internal sulfate attack; see Figure 4.2). There is at least one reported case where similar map cracking has been noticed in the absence of any alkali-reactive aggregate in the concrete, but where internal sulfate attack was clearly demonstrated. These observations are not surprising, as in all cases of volumetric expansion the cracks will always be preferentially parallel to the stresses.

### Microscopic observations

Most of the available microscopic information has been obtained using light optical microscopy and scanning electron microscopy (SEM) in back-scattered

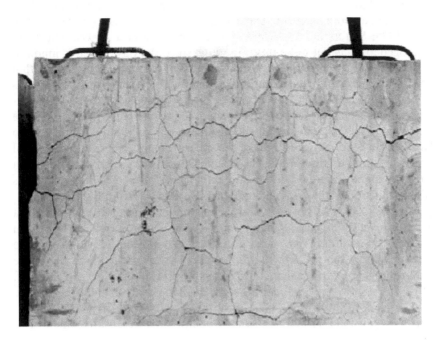

*Figure 4.2* Prefabricated concrete beam damaged by combined action of ASR (primary mechanism) and heat-induced ettringite expansion (Photo courtesy of M. Thomas).

electron (BSE) mode. Reported microscopic observations of concrete damage include:

1 expansion of the cement paste as revealed by formation of partial or complete rims (gaps, bands, circumferential cracks), up to about 25–30 micrometers wide, around the aggregate particles (e.g. Johansen *et al.* 1993);
2 partial or complete filling of these gaps by secondary ettringite;
3 formation of "nests" of ettringite in the cement paste (e.g. Marusin 1993);
4 formation of two-tone C-S-H features (e.g. Scrivener 1992); and
5 microcracking of the paste.

Most important, it is now possible to perform

6 microanalytical studies on the composition of hydration products, thus develop better understanding related to the physico-chemical processes at the paste micro-scale (e.g. Scrivener and Taylor 1993; Famy 1999).

One of the characteristic features of heat-induced sulfate attack is the formation of circular or peripheral cracks (gaps, bands), around the aggregate particles. There is some evidence to show that the width of these gaps is proportional to the aggregate size, this fact implying that the formation of the gaps is related to homogeneous expansion of the cement paste (Johansen *et al.* 1993, 1995; Skalny *et al.* 1996; Johansen and Thaulow 1999). The term "homogeneous" is not to be taken literally, since the paste itself is inhomogeneous (e.g. local variations in paste density) and the micro-conditions within the paste clearly vary from site to site; thus the experimental measurements may not be perfect in all cases. These local variations obviously lead to variations in expansion and crack density. Data of Odler and Chen (1995, 1996) are in support of the phenomenon of "homogeneous" expansion. Please note also that the thin sections used to make the measurements are two-dimensional representations of three-dimensional particles; because of this, for each particle diameter there is a lower boundary value for the gap width, which is a linear function of the particle diameter (see data points in Figure 4.3).

Needless to say that not everybody is in complete agreement with the idea of paste expansion (e.g. Diamond *et al.* 1998; Diamond 1996b; Zhaozhu Zhang 1999), and data have been presented to show that it is not the paste expansion *per se* but the growth of ettringite crystals in the paste and at the paste-aggregate interface that control the overall expansion (e.g. Yang *et al.* 1999). In other words, the claim has been made that secondary ettringite does not deposit into empty spaces generated by the paste expansion (whatever the expansion mechanism may have been), but is at least to some degree

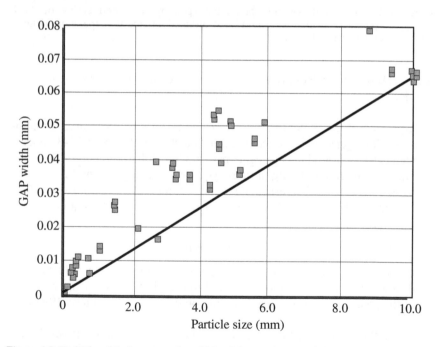

*Figure 4.3* Relationship between the width of the gaps around the aggregate particles as a function of the aggregate size (Photo courtesy of N. Thaulow).

the *cause* of paste expansion. More about mechanistic aspects of heat-induced sulfate attack later.

According to Erlin (1996), and in agreement with other literature, expansion of the paste and the consequent peripheral paste-aggregate separations may also be caused by other mechanisms, such as freezing and thawing and expansion in paste of free lime or free magnesia. Thus, detection of ettringite-filled or partially-filled gaps around the aggregate particles is not an adequate characteristics of DEF. Gaps around aggregate particles may occur even in the absence of ettrigite in them. An example of concrete showing empty gaps around the aggregate periphery is shown in Figure 4.4.

According to Yang and coworkers (1999a), in the absence of heat-curing, visible ettringite in the paste is seldom identified; most of the detected calcium sulfoaluminate is present in the form of calcium monosulfoaluminate hydrate, or monosulfate. In heat-cured samples no ettringite could be found immediately following the heating, and the ettringite morphologies in the paste and at interfaces were the same as in samples cured at ambient temperature. Ettringite could initially be observed only after about ninety days of water storage at room temperature, both in the paste matrix and at the interfaces. After about 155 days of water curing, even better developed ettringite bands were found, especially in the vicinity of and parallel to the surfaces of the test samples. The bands at the interfaces were seldom observed to fully fill the circumferential cracks at the interfaces. Please note that the expansion

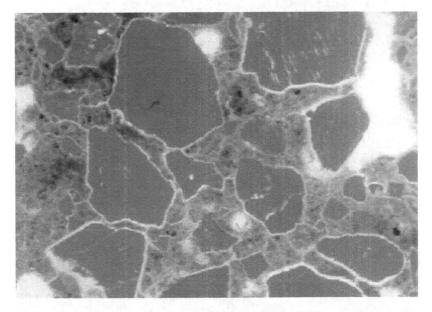

*Figure 4.4* Empty gaps around aggregate particles in a concrete railroad tie from Scandinavia. Field of view: 6.3 × 4.1 mm, optical light microscopy: fluorescent light (Courtesy of U. Hjorth Jakobsen).

effect is sample-size dependent, thus comparison of laboratory samples to field concrete is complex.

An example of ettringite deposited into the pre-existing C-S-H ("nests" of secondary ettringite in paste) is presented in Figure 4.5. Areas of such ettringite deposition are often accompanied by local expansion and by development of a network of microcracks. The morphology of such ettringite-rich paste seems to be identical in both internal and external forms of sulfate attack.

The effect of temperature on the kinetics of hydration of cement and its components is well known, as is the effect of curing conditions on the mechanical properties of concrete. It is less appreciated, however, that temperature also affects the microstructure and crystal habit of the hydrates formed during hydration and that these effects, in turn, are causing, at least to some degree, the observed physical changes. Examples of temperature-induced changes are partial or complete decomposition of hydration products, change in molar composition of the hydrates (e.g. the calcium–silica ratio, C/S), changing reactivity, weakening of paste to aggregate bond, etc.

Availability of improved techniques of sample preparation and modern computerized scanning instrumentation allows now detection of subtle variations in the microstructure of hydration product and, more important, enables to relate these observations to concrete processing practices. Example of such relation-

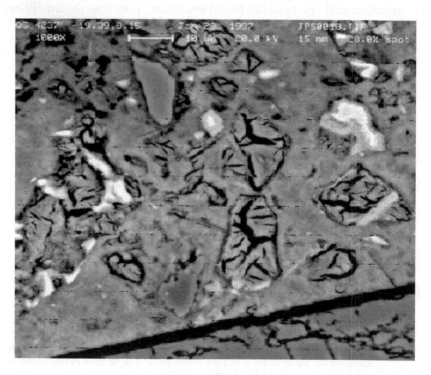

*Figure 4.5* Secondary ettringite "nests" dominating the local paste morphology (Photo: J. Skalny).

ship is the observation that certain microscopic features appear in hydrated cement paste exclusively when the paste or concrete has been cured at elevated temperatures. One feature believed to be typical for concrete exposed to elevated temperature is the so-called "two-tone" C-S-H, also referred to as high-temperature "relics" or "dense shells surrounding the cement grains," etc.

The "two-tone" C-S-H feature was first reported in the late 1980s and early 1990s (e.g. Thaulow 1987; Scrivener 1992; Clark *et al.* 1992; Jie *et al.* 1993; Famy 1999). At elevated temperatures, such as during steam curing, the decreased solubility of calcium hydroxide and rapid hydration are believed to lead to formation of dense hydration rings around the calcium silicate (alite, belite) particles. As a result, especially at longer hydration times, their degree of hydration decreases, this in conformity with the observation that pastes cured at elevated temperatures have a lower degree of hydration and lesser amount of formed calcium hydroxide than pastes cured under ambient conditions (Clark *et al.* 1992; Jie *et al.* 1993). Upon conclusion of heat curing, the concrete continues to be cured at ambient temperatures,

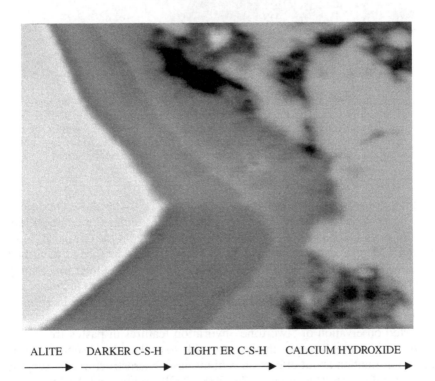

ALITE    DARKER C-S-H    LIGHT ER C-S-H    CALCIUM HYDROXIDE

*Figure 4.6* Two-tone C-S-H in an incompletely hydrated mortar sample precured for four hours at 20 °C, heated at 90 °C for twelve hours and subsequently stored in water for 400 days. High magnification BSE image. Note that the alite core (left) has not completely reacted and is rimmed with two-tone inner C-S-H. The darker inner C-S-H rim grows thicker as alite continues to react.

*Source*: Famy (1999)

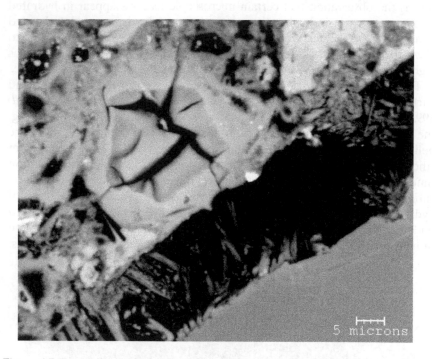

*Figure 4.7* Internally cracked fully hydrated cement grain exhibiting the two-tone C-S-H rims. The microcracks have started from the darker C-S-H core and spread further in the lighter rim. Ettringite is observed within the gap at the paste-aggregate interfaces. Mortar sample precured for four hours at 20 °C, heated at 90 °C for twelve hours and subsequently stored in water for 600 days. High magnification BSE image.
*Source*: Famy (1999)

and the newly formed hydration products produced under these conditions may have somewhat different density and, possibly, molar C/S composition. (see Figures 4.6 and 4.7). Such effect of temperature on the composition of C-S-H, including its surface area and C/S and H/S ratios, has been established years ago (e.g. Kantro *et al.* 1963; Skalny and Odler 1972; Odler and Skalny 1973). These differences in density (and possibly composition) are visible in backscattered mode of SEM, thus were proposed as qualitative "fingerprints" in estimating the processing temperature. In 1998, Diamond *et al.* (1998) reported observation of "two-tone" hydration features in pastes treated at *room* temperature; however, they were unable to explain the observation. For more information about effects on cement paste microstructure, we suggest reading references (Diamond 1976, 2000, 2001; Gollop and Taylor 1992–1996, Scrivener 1992, Damond and Lee 1999, Famy 1999).

Scrivener (1992), Scrivener and Taylor (1993), Lewis (1996), Famy (1999), and Taylor *et al.* (2001) published microanalytical data on the composition of C-S-H formed under conditions of heat-induced sulfate attack. Using back-scattered electron images (BSE), in combination with image analysis, it is possible to distinguish the most important constituents of a cementitious system by their distinct backscatter coefficients and grey levels of their images. Anhydrous compounds typically appear to be the brightest, empty pores appear to be the darkest; the grey level of hydrates is located in between. (see Figure 4.8).

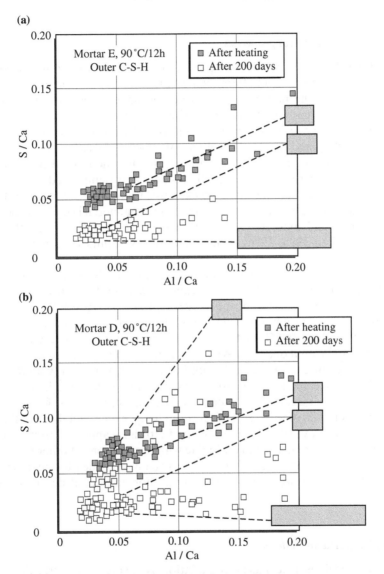

*Figure 4.8* An example of the use of EDS microanalysis (determination of atom ratios of S/Ca versus Al/Ca) as a tool for mechanistic understanding of delayed formation of ettringite. Compositions of two mortars heated at 90 °C immediately after heating and after 200 days storage in water: (a) expansive mortar; (b) non-expansive mortar.

*Source*: Famy (1999)

In combination with other analytical data, such microscopic information can be used in identification of the concrete damage mechanisms. Needless to say this approach is only qualitative, or at best semi-quantitative and, at least at the present time, it is primarily used to support or confirm data obtained by more conventional quantitative techniques.

### Deterioration of physical properties

The in-depth studied physical properties of concrete exposed to excessive heating are: paste expansion, formation of micro-cracks, and the related over-all expansion of the mortar and concrete specimens. The paste expansion and formation of gaps has been briefly discussed above.

As the paste expansion in concrete increases, the formed micro-cracks are inter-connecting and the overall volume of concrete increases. It should be noted, than not all cement compositions expand when cured at elevated temperatures. Lawrence, Siedel and their coworkers published data showing that heat-induced internal sulfate attack acting in the absence of other damage mechanism might cause expansion in some commercial cement products cured at 80 °C and above, but not in all (Siedel *et al.* 1993; Lawrence 1995a,b; Odler and Chen 1995, 1996). Similar laboratory results were reported by others.

Recently published data on laboratory mortar prisms by Yang *et al.* (1999b) clearly demonstrate the differences in expansion and rate of ettringite development between heat-cured and control prisms (Figure 4.9). Whereas control samples did not show any significant expansion, expansion of the samples heated at 100 °C for twelve hours initiated within ninety days and was completed within one year. Whereas in room temperature cured samples most of the ettringite formed at the initial stages of hydration (within seven days), in heat-treated samples ettringite could not be detected immediately following the heat treatment. Detectable ettringite started to be formed only at about seven days of water storage at room temperature and then continued to increase. The overall amount of formed ettringite appeared to be always lower in the heat-treated samples. In line with observations of other researchers, formation of ettringite after cooling usually precedes the measurable expansion by many weeks or even months. For additional discussion see Chapter 5.

### Examples of structural damage

The most obvious mechanical degradation is severe macro-cracking and large-scale expansion that may, in extreme cases, lead to loss of concrete strength and complete structural failure. Development of severe cracking and expansion has been observed in railroad ties (e.g. Figures 1.3, 8.10), prefabricated structural elements (e.g. Figure 4.2), and small-size heat-cured

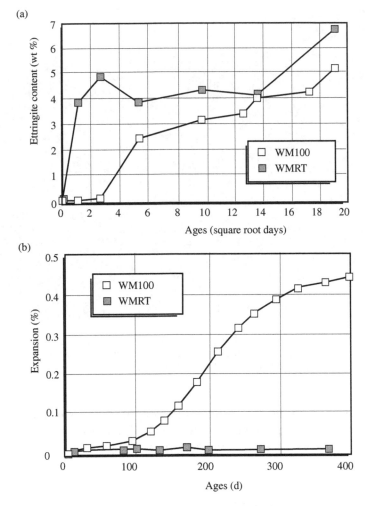

*Figure 4.9* Ettringite contents (a) and expansion (b) of mortars cured at room tem-
perature (WMRT) and at 100 °C (WM100).
*Source*: Reprinted from *Cement and Concrete Research*, vol. 29, Yang R., Lawrence, C.D.,
Lynsdale, C.J. and Sharp, J.H: "Delayed ettringite formation in heat-cured Port-
land cement mortars", pp. 17–25, Copyright 1999, with permission from Elsevier
Science

concrete products (e.g. Figure 4.10) (Scrivener 1996; Lawrence *et al.* 1999;
Thomas 1998). There is no credible evidence for occurrence of so-called
DEF in concrete produced at ambient temperature. Again, clear-cut
description of heat-induced internal sulfate-related mechanical damage is
difficult, as most known cases showed a combination of sulfate damage
mechanisms with ASR and, possibly, freezing–thawing. In contrast with

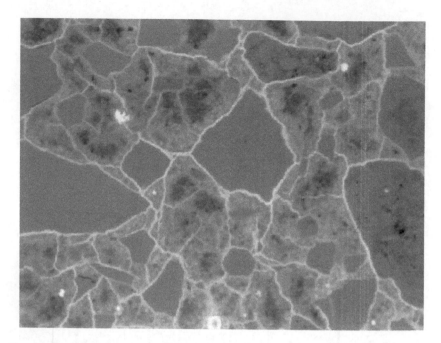

*Figure 4.10* Empty gaps around aggregate particles in a defective roof tile product. Light optical microscopy: fluorescent light, field of view: 6.3 × 4.1 mm (Photo courtesy of U. Hjorth Jakobsen).

some claims are difficulties of identification, the authors believe that easy identification and separation of ASR and DEF is possible by a combination of optical and electron-optical microscopic techniques.

### Discussion of controversial DEF issues

*Suggested mechanisms.*   It is clear from the above discussion that the mechanistic processes leading to heat-induced internal sulfate attack are complicated, thus confusing. Whereas some believe that the concrete expansion is caused by delayed formation of ettringite in the paste and its subsequent expansion and cracking (e.g. Johansen *et al.* 1993), others claim the secondary ettringite deposited or reprecipitated within the cracks, air voids and gaps, to be expansive (e.g. Heinz 1986), or attribute the damage to the effect of nucleation of ettringite in the crack-tip zone (Fu *et al.* 1993, 1994; Fu and Beaudoln 1996), or to the hydraulic pressure originating in osmosis (Mielenz *et al.* 1995). There are several other "variations on the above tune". The basic ideas are as follows.

Ettringite formation is a normal and anticipated process of Portland cement hydration and, under regular circumstances, it does not lead to expansion. It is common to find it in cracks and voids of matured and aged concrete exposed to varying climatic conditions. This is the result of recrystallization, also known as "Ostwald ripening." The smaller crystals are thermodynamically less stable than large crystals. Under moist conditions the smaller ettringite crystals dissolve in pore liquid and recrystallize as larger crystals in the available spaces. In concrete damaged by excessive heating, alkali silica reaction, or other mechanisms, the moisture content is typically higher and changes more frequently than in sound concrete. Consequently, the recrystallization is more efficient. In this respect the behavior of ettringite and calcium hydroxide are similar.

The formation of ettringite is not always accompanied by expansion, and the expansion is not always proportional to the amount of ettringite formed. Ettringite is not expansive *per se*. It is the conditions under which the ettringite forms that cause or control expansion of mortars and concretes.

According to Heinz and Ludwig (e.g. Heinz 1986; Heinz and Ludwig 1986), the origin of the degradation is the formation of ettringite at the cement-aggregate interface. Early experiments supported this argument, in that experimental data have shown that neat cement pastes do not expand whereas mortars and concrete, containing aggregate, do expand (e.g. Lawrence 1993a,b). However, new experimental data show that paste specimens expand too, but the time needed to the onset of expansion may be greatly delayed (an induction period) and the ultimate expansion is lesser than in samples of concrete (e.g. Odler and Chen 1995, 1996; Lawrence 1995a,b).

According to Johansen *et al.* (1993), Taylor (1997) and others, the expansion is preconditioned by (a) thermal decomposition of the primary ettringite formed at early stages of hydration (or its inability to form at elevated temperatures); and (b) the renewed availability of sulfate and aluminate ions after cooling to enable formation of ettringite intimately dispersed within the C-S-H. Such paste expands (no mechanism of paste expansion given), leading to formation of gaps around the aggregate that are detectable by microscopic techniques. The intimately dispersed ettringite eventually recrystallizes into available space by the "Ostwald" mechanism discussed above (forming secondary ettringite). This process is believed to be non-expansive. Johansen *et al.* (1993) explain the mechanism, using calcium hydroxide formation as an analogy, as follows:

The formation of $Ca(OH)_2$ in voids and cracks in hardened cement paste and mortars is not considered to cause expansion, in contrast with the possible effect of the formation of $Ca(OH)_2$ from CaO (or $Mg(OH)_2$ from MgO). In a cement paste saturated with water, gradients for $Ca(OH)_2$ are set up. The small $Ca(OH)_2$ crystals formed during the hydration of the clinker minerals and dispersed throughout the system have a slightly higher solubility relatively to the large crystals of

Ca(OH)$_2$; eventually larger crystals will grow in the voids and cracks while the small crystals disappear (Ostwald ripening).

There may be another effect due to wetting and drying, resulting in a repeated dissolution and precipitation of Ca(OH)$_2$ in the voids and cracks. This effect may be larger than the Ostwald ripening. Recrystallization of ettringite takes place in a similar way

(Johansen *et al.* 1993).

The time dependence of the paste expansion is presented in micrographs of Figure 4.11 (Pade *et al.* 1997). It clearly demonstrates changes in microstructure with time, particularly the frequency of crack formation and crack width growth.

In a series of articles, Fu *et al.* (e.g. 1993, 1994) and Fu and Beaudoin (1996) have suggested that heat-induced expansion may be preconditioned by prior formation of ettringite nuclei in the paste microcracks, specifically at the crack tips. In their interpretation, excessive curing temperature and, possibly, the accompanying or subsequent high-temperature drying, lead to rapid adsorption of sulfate by C-S-H, followed by its slow release upon cooling

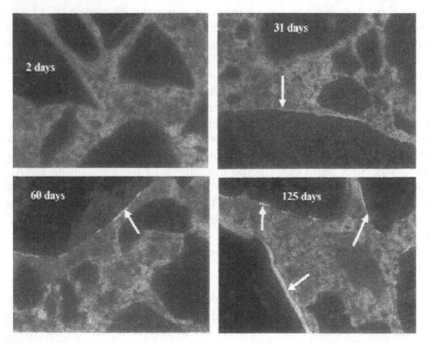

*Figure 4.11* Thin sections of heat-treated mortars at various ages: crack pattern and crack width as a function of time. Light optical microscopy: fluorescent light, field of view: 0.5 × 0.7 mm (Courtesy of U. Hjorth Jakobsen)
*Source:* Pade *et al.* (1997)

and subsequent moist curing into the liquid phase. Once supersaturation of the liquid phase with respect to ettringite formation is reached, nuclei of ettringite form by a through-solution mechanism, preferably in preexisting cracks. The size of these cracks may be critical; larger preexisting cracks will lead to greater expansion. The growth of the ettringite nuclei in limited space, results in crystallization pressure and subsequent extension of the cracks.

According to the above authors, the primary source of preexisting cracks is shrinkage, especially repeated drying shrinkage. It is speculated that drying shrinkage may lead to release of sulfate from C-S-H. The sensitivity of concrete to steam curing seems then to be related to at least these phenomena: (a) decomposition or non-formation of ettringite due to high temperature; (b) adsorption of the released sulfate by C-S-H; (c) possible formation of microcracks due to thermal expansion; (d) release of sulfate from C-S-H upon cooling and subsequent ambient temperature moist curing; (e) formation of microcracks as a result of drying shrinkage; (f) formation of ettringite nuclei in the preexisting cracks, and (g) growth of the nuclei resulting in paste expansion.

It is not clear how the release of sulfate from C-S-H can lead to supersaturation with respect to formation of ettringite nuclei. The usual, unconvincing explanation is that it can occur locally or on micro-level. The need for preexisting cracks in heat-induced internal sulfate attack, formed as a consequence of drying shrinkage or mechanisms of deterioration such as freezing–thawing, alkali–silica reaction-related or thermal expansion, is argued also by Sylla (1988), Johansen *et al.* (1993), Skalny *et al.* (1996), Taylor (1997) and others. In other words, it seems that the expansion caused by DEF in itself is a weak effect, an effect that would in the absence of preexisting cracks caused by other phenomena not cause measurable damage in most cases.

Mielenz *et al.* (1995) published their interpretation of the observations collected during a litigation involving precast concrete railroad ties (see also Chapter 8, Case Histories). The two technical issues of importance in this litigation were (a) the role of ASR versus sulfate attack; and (b) the source of sulfate (high-sulfur clinker/cement or heat-induced decomposition followed by recrystallization of primary ettringite). In view of the above authors, the observed expansion is the result of formation in the paste of an amorphous-looking substance that is either a colloidal mixture containing sub-microscopic ettringite crystals, or a dispersed solid similar in chemical composition to ettringite, or a combination of such solids:

> The gelatinous-appearing substance seemed to have been generated within the matrix as a viscous gel or colloidal sol, and to have swelled or increased in volume, probably by osmosis with uptake of water through the cement paste, in which the calcium silicate hydrates acted as a semi-permeable membrane. As the uptake of water continued within the critically saturated matrix, the gelatinous substance was initially accommodated by intrusion into

previously existing, air-filled space, such as air voids, cracks within aggregate particles, and separations at the perimeter of aggregate particles, the latter being a consequence of bleeding and settlement of the fresh concrete.

The presence or absence and thickness of the linings were indifferent to the mineralogy of the particles, indicating that this substance was not a product of a cement-aggregate reaction.

...the DEF was caused by the presence of large amounts of sulfate in the cement (probably in relation to the $C_3A$ content) and by the very slow solubility of much of that sulfate due to its presence in clinker phase of cement

(Mielenz *et al.* 1996).

This interpretation is clearly in conflict with those presented earlier. It implies the existence of a colloidal form of ettringite, attributes the formation of gaps (bands, linings, peripheral cracks) to bleeding or settlement of the concrete, introduces "osmosis with uptake of water" as the damage mechanism, and alleges the clinker sulfate to be the cause of "DEF." This conclusion seems to be in conflict with some of the experimental data obtained by the authors themselves and by others, showing the damage to be caused by a combination of ASR and heat-induced, rather than excessive clinker sulfate-induced, internal sulfate attack.

Wieker and Herr (1989) and Brown and Bothe (1993), discuss the effect of alkali content on the mechanism of sulfate attack. They show that the concentration of sulfate ions in the pore solution increases with alkali content and temperature, and this increased sulfate concentration may be related to the observed greater expansion of such mortars and concretes upon cooling. The presence of alkalis is most probably also related to the observed simultaneous occurrence of ASR, "DEF" and possibly, could explain the fact that "DEF" is almost always accompanied by ASR. More about this phenomenon will be discussed in this chapter.

Based on theoretical and experimental data in the literature, Taylor (1994b, 1996) suggested a tentative mechanism for cement mortar assuming a reaction front advancing into the material (see Figure 4.12):

The first thing that happens is that (primary) ettringite is formed as microcrystals dispersed in the C-S-H gel. These tend to recystallize (*as secondary ettringite crystals*) in Hadley grains and other existing cavities. The paste expands, due not to this recrystallization but the processes occurring on a smaller scale, and this expansion extends and widens the cracks, especially around the aggregate grains. Because the expansion is not uniform, larger cracks also form in the paste. Ettringite recrystallizes in the cracks, perhaps as quickly as the space becomes available. If the bond between paste and aggregate is sufficiently strong, the paste cannot expand, because it is restrained by the aggregate

(Taylor 1994b).

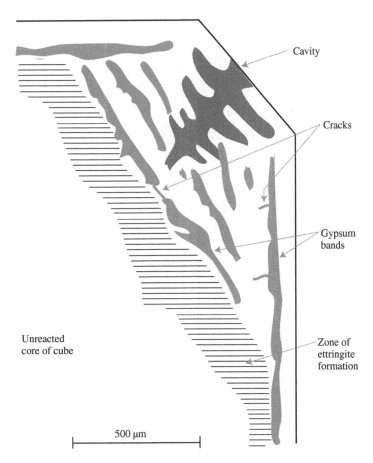

*Figure 4.12* Suggested mechanism of expansion in mortar.
*Source*: Reprinted from *Cement and Concrete Research*, vol. 22, Gollop, R.S and Taylor, H.F.W, "Microstructural and microanalytical studies of sulfate attack in ordinary cement paste", pp. 1027–1038, Copyright 1999, with permission from Elsevier Science

In a series of recent presentations and publications, Famy, Scrivener, and Taylor updated the above mechanism and proposed the following sequence of processes causing *delayed* ettringite-related damage (Famy 1999; Taylor 2000a, 2000b, 2000c, Taylor *et al.* 2001). Following partial or complete thermal decomposition of primary ettringite during the curing process, the subsequent cooling to ambient temperature enables formation of sub-micrometer, primary (!) ettringite crystallites in the *outer* product C-S-H (see Figure 4.13). The principal reactants in formation of this ettringite are monosulfate, C-S-H, and the pore solution. The role of alkalis in the pore solution on the extent of expansion, if any, is not entirely clear.

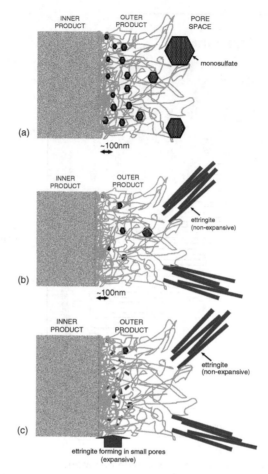

*Figure 4.13* Formation of expansive forces after heating: (a) microstructure immediately after heat curing; (b) microstructure after subsequent water storage for a non-expansive mortar; (c) microstructure after water storage for an expanded mortar.

*Source*: Famy (1999)

The ettringite crystallites form in the close vicinity of the reacting monosulfate particles. With increasing amount of newly formed ettringite, an uneven expansion of the paste begins and cracks develop in the paste and at the paste-aggregate interface (see Figure 4.14). The expansive forces are decreasing with increasing distance from the original grain boundary (boundary between *inner* and *outer* product C-S-H), most probably because of the increasing porosity of the *outer* product C-S-H. Due to Ostwald ripening the formed ettringite particles slowly recrystallize into both pre-existing and paste expansion-caused open spaces of all kinds (air voids, Hadley grains,

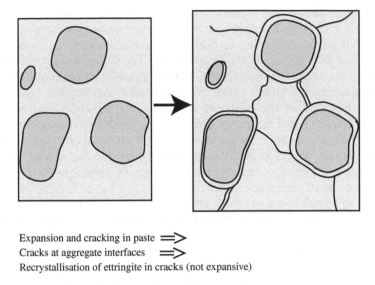

Expansion and cracking in paste ⟹
Cracks at aggregate interfaces ⟹
Recrystallisation of ettringite in cracks (not expansive)

*Figure 4.14* Paste expansion and formation of cracks at interfaces and paste.
*Source*: Famy *et al.* (2001)

gaps around aggregates, etc.) as *secondary* ettringite. All these processes are believed to occur simultaneously rather than sequentially.

As of today, the exact relationship between the amount and rate of renewed ettringite formation on the one hand, and the degree and rate of expansion on the other hand, is not entirely clear. Neither is it obvious why some cement compositions are more prone to expand than others; many of the available correlations are inconclusive. Nevertheless, serious progress to explain the heat-induced sulfate expansion and related phenomena has been made, and some of the issues will be discussed in more detail in the following paragraphs.

*Effect of temperature and other curing conditions.* It is now accepted that excursion of the concrete curing temperatures to above 65–70 °C may lead to the heat-induced ("DEF") problems highlighted above. It is unresolved, however, whether the same phenomenon can occur at ambient temperature.

Most experimental and field studies show that ettringite cannot be detected immediately after high-temperature curing, suggesting that ettringite is unstable at temperatures above about 65–70 °C. However, based on studies by Heinz and Ludwig (1986, 1987), Wieker and Herr (1989), Brown and Bothe (1993), Glasser (1996), Batic *et al.* (2000) and others, it seems that the situation is more complicated and is related to the ionic composition of the liquid phase in equilibrium with the solids at the given temperature, and in particular to the solubility of sulfates. It has been shown now that under favorable conditions ettringite may be stable even at 100 °C or higher. The distribution of sulfate and aluminate between the pore solution and solids, particularly C-S-H, will depend on their concentration, temperature, and presence of alkali. According to Scrivener and coworkers (Scrivener and Taylor 1993; Lewis *et al.* 1995; Taylor *et al.* 2001), the amount of sulfate and aluminate in the inner product C-S-H increases with temperature, and the sulfur–aluminum atomic ratio, taken immediately after the heat treatment, is indicative of the potential for future expansion. Upon cooling, ettringite is formed and this newly formed ettringite is associated with paste expansion discussed above. Such post-curing formation of ettringite is, however, not directly related to the total amount of ettringite that can be formed, and there is some evidence that the ettringite (e.g. detectable by XRD) is not the direct cause of expansion.

Claims of "ambient-temperature DEF" were originally reported by Hime (1996a, 1996b) and Marusin and Hime (1996), later also by Collepardi (1999a) and Lawrence *et al.* (1999). Unfortunately, the presented experimental data are incomplete in that they do not give adequate information about concrete composition (incl. sulfate form and content in the clinker) and processing (curing conditions, ambient temperature, internal heat of hydration), thus do not provide credible evidence of ettringite expansion-related damage at low temperatures. Similar arguments were presented by Diamond (1996a, 1996b), referring to one particular case. Again, the limited data presented are inconclusive, as there is unpublished evidence showing that the concrete in question has been heat-treated (although the exact temperature is unknown) and the curing practices may not have been in line with best practices of concrete making, or both. It should be noted that, with the exception of the case reported by Diamond and a few other cases, all alleged claims of concrete damage due to "ambient-temperature DEF" were complicated by the fact that ASR-related phenomena were detected in the same samples. More about combined ASR and heat-induced internal sulfate attack later.

According to Thomas (1998, 2000), the root of the "low-temperature DEF" idea is the above-mentioned railroad tie litigation in the early 1990s, in which it was alleged that concrete sleepers produced on Fridays were showing damage in spite of the fact they were not steam cured; if correct, this would support the allegation that ambient temperature "DEF" is possible. However, it now looks this may not have been the case because, based on documentation available from the trial, some of the "Friday ties" were actually steam cured, and those steam cured "Friday ties" were the ties reported to be damaged (see also Chapter 8 on Case Histories).

According to Taylor (1997, 2000a), there is no credible diagnosis that would suggest existence of "DEF" in concrete not deliberately cured at an elevated temperature. However, as also discussed by others, he cites the possibility that high heat of hydration and exposure to sunlight, or both, may aid in increasing the temperature to levels that may cause decomposition of primary ettringite formed during the early stage of hydration.

Another poorly-substantiated "ambient-temperature" ettringite-related damage is the allegation that concrete pavements may fail prematurely due to in-filling of air voids by (delayed?) ettringite, thereby reducing the effectiveness of the air void system (e.g. Marks and Dubberke 1996; Quyang and Lane 1997; Stark and Bollmann 1997, 2000). The available data are inconclusive, especially in view of the data published by Detwiler and Powers-Couche (1996), who concluded that any ettringite deposition in the entrained air voids is the consequence of the damage to the paste pore structure caused by repeated freezing–thawing. Similar conclusions were reached by Mather (1993) and Skalny (1994).

*Role of clinker and cement composition.*  It has been alleged that clinker or cement composition, in particular the form and amount of sulfate present, are responsible for at least some of the expansion and cracking associated with the DEF process (e.g. Mielenz *et al.* 1995; Hime 1996a; Collepardi 1997, 1999b). These claims are based on early DEF-related data by Heinz, Ludwig and others, which suggested that expansion in mortars and concretes increases with increasing $SO_3/Al_2O_3$ (Heinz *et al.* 1989; Heinz and Ludwig 1987) or that there exists a pessimum $SO_3/Al_2O_3$ mass ratio of ca 0.8 (Ludwig 1991). These reports were later complemented by claims that there is a pessimum Bogue $C_3A$ content (Grattan-Bellew *et al.* 1998) or that expansion may be related to $C_3A$ content and that it is highest at ca 4% of $SO_3$ (Kelham 1999). The issue of clinker sulfates was also studied by Odler and Chen (1995), who reported data for four clinkers having different $SO_3/Al_2O_3$ mass ratios. Significant expansions were found only for clinkers with high $SO_3$ *and* $C_3A$ contents, but only when they were heat-cured at 90 °C. Miller and Tang (1996), Klemm and Miller (1997), Skalny *et al.* (1997), Michaud and Suderman (1997), Tennis *et al.* (1997), Montani (1997) and others reported data on sulfate analyses of industrial clinkers. All these studies concluded that there is no experimental laboratory evidence that sulfate present in clinker, in whatever form, that might undergo "late" release into hydrating system, can lead to expansion in non-heat cured mortars or concretes.

The role of cement alkalis is equally unclear. According to Wieker and Herr (1989), expansion increases with the cement sodium equivalent, $Na_2Oe$ ($Na_2O + 0.66K_2O$), but alkali in their cements correlated with the $SO_3/Al_2O_3$ ratio. More recent work correlates the alkali content with increased sulfate levels (Kelham 1996, 1999; Famy 1999). According to Hime (1996a), an imbalance between sulfates and alkalis may cause sulfates to be incorporated in clinker calcium silicates (belite, alite), but no experimental evidence to support these claims is given.

Lawrence (1995b, 1999) and Kelham (1999) also reported that there might be correlations between expansion and MgO content, cement fineness, and calculated $C_3S$ content. Lawrence (1999), Kelham (1999), and Hobbs (1999) performed multiple regression analyses, but the results reveal a high complexity of the relationship. According to Taylor (2000b) and Taylor *et al.* (2001), the obtained relationships are most complex and, at present time, inadequate to predict expansion based on clinker or cement compositions.

According to Thomas (1998, 2000), the evidence in the literature shows clearly that cements produced from high-reactivity clinkers – i.e. clinkers with high $C_3S$, $C_3A$ and alkali and sulfate contents – ground to high specific surface areas, have the greatest susceptibility for expansion due to *delayed* ettringite formation...but only when heat cured. Such high-reactivity clinkers are typical for high-early strength cements and Kelham's correlation between the two-day compressive strength of mortars and their expansion clearly confirms such relationship (1996). This implies that use of reactive cement under conditions where heat-curing is used for additional acceleration of early strengths, should be considered with caution. In line with others looking into this issue, Thomas did not find any evidence of ambient-temperature DEF.

*Role of aggregate type.* The degree of heat-induced expansion of mortars and concretes was observed by several authors to depend on the type of aggregate. According to Yang *et al.* (1999b), there is a clear difference between the effect of various types of aggregate and, in products that were made with lower sulfate cement or higher w/cm, some aggregates might even prevent the expansion entirely. For example, a limestone aggregate was found to greatly delay the onset and reduce the extent of the expansion in both mortar and concrete. In contrast, heat-cured granite concrete showed early and large expansion upon cooling and moist curing, and heat-cured flint concrete gave an intermediate induction period and ultimate expansion.

It has also been reported by the above authors that formation of ettringite in the gaps around the aggregate and in the paste is, in some non-quantifiable manner, related to the aggregate quality. In heat-cured quartz mortars, ettringite bands developed mostly around the sand grains. For mixed quartz and limestone mortars, the bands formed primarily around the quartz sand grains. For granite as well as flint based concrete, ettringite bands developed around many aggregate particles and seemed to be interconnected through the paste matrix (by ettringite morphologies described earlier as ettringite "nests"). Finally, in heat-cured limestone concrete, the number of ettringite bands around the particles was significantly reduced and many ettringite morphologies developed strait through the paste matrix. The quartz sand mortars exhibited large expansions within the first year, but the limestone mortars remained stable even after six years. The ultimate expansion of mortars made with a combination of limestone and quartz aggregate had intermediate values.

Based on microstructural observations, the above authors believe that expansion of heat-treated mortars and concrete is caused by formation of ettringite bands, and is unrelated to the quasi-*homogeneous* expansion of the concrete matrix claimed by Johansen *et al.* (1993). The relevance of the above argument is unclear, however, as the proponents of paste expansion do not relate the expansion to the visible ettringite observed by Yang and coworkers, but to the invisible, probably microcrystalline, ettringite formed by renewed availability of needed ionic species released upon cooling of the system. Such ettringite is believed to be "primary" ettringite that with time recrystallizes and grows into larger, secondary ettringite particles (bands, nests), such as are visible in backscattered mode by SEM. According to Yang and coworkers, the reduced expansion observed in limestone concrete might be the result of the rough limestone aggregate surface, resulting in enhancement of the paste-limestone bond and interference with the development of expansion-causing ettringite bands.

The effect of aggregate on expansion was also studied by Grattan-Bellew *et al.* (1998), who found the expansion and the amount of ettringite formed to be inversely proportional to the particle size of the quartz aggregate, the only aggregate that caused serious problems. This effect is believed to be related to the coefficient of thermal expansion of the particular aggregate; when the coefficient is larger than about $10 \times 10^{-6}/°C$, upon thermal cycling such mortar undergoes sufficient expansion to create microcracks and crystallization sites for ettringite nuclei. The other aggregates tried were basalt, dolomitic limestone (dolostone), granite, limestone, and siliceous limestone. Note that in contrast to these results, Yang *et al.* (1999) reported extensive expansion in granite-containing concrete.

*Effect of other simultaneous mechanisms of deterioration.* Based on laboratory studies and field experience, it seems that the probability of damage to concrete by delayed formation of ettringite increases for concrete that was simultaneously or previously damaged by other possible mechanisms. Such damage mechanisms include thermal cracking, alkali–silica reaction and, possibly, repeated changes in temperature and humidity (freezing and thawing, drying and wetting). All DEF cases, with one above-noted questionable exception, are known to be concrete products or structures exposed to high curing temperatures in the form of steam curing or a combination of high ambient temperature and excessive heat of hydration (highly-reactive cements, high surface area). Although almost certain, this view is based on circumstantial rather than solid experimental evidence, and a well-designed and executed study is needed to resolve this controversial question.

In the majority of reported cases, DEF occurred simultaneously with ASR (e.g. Shayan and Quick 1991, 1992; Oberholster *et al.* 1992; Johansen *et al.* 1993; Lawrence *et al.* 1999; Thomas 1998). This is believed to be related to the continuous decrease during alkali–silica reaction of the alkali concentration of the pore solution: alkali is used up in formation of calcium–alkali–silica gel, the pore solution alkali concentration decreases, and the lowered alkali concentration enables increased rate of ettringite formation. The reasons for preferential formation of ettringite at lower alkali concentrations were discussed by Brown and Bothe (1993), Herr *et al.* (1988), Wieker and Herr (1989) and others.

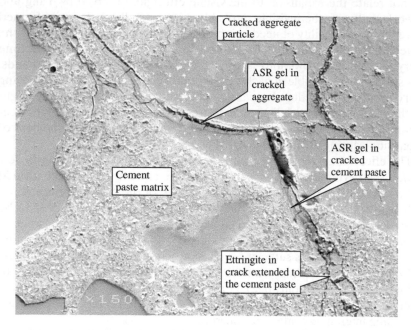

Cracked aggregate particle

ASR gel in cracked aggregate

ASR gel in cracked cement paste

Cement paste matrix

Ettringite in crack extended to the cement paste

*Figure 4.15* Micrograph showing cracking of an aggregate particle caused by formation of ASR gel within the aggregate and extending the crack into the paste. The crack is subsequently filled with secondary ettringite. SEM, backscattered mode (Photo courtesy of Thomas and Ramlochan, University of Toronto).

Although doubted by some authors, evidence is mounting that, in cases where combined ASR and DEF mechanisms are operable, ASR is the primary mechanism of damage. It seems now possible that in the absence of ASR or other mechanisms of expansion and cracking, delayed formation of ettringite by itself would not have caused damage in most cases. The micrograph shown in Figure 4.15 represents a concrete sample taken from a structure diagnosed as showing signs of combined ASR and internal sulfate attack. It clearly indicates the primary formation of ASR gel, followed by ettringite in-filling of the created crack.

### 4.8.3   Concluding comments on *internal* sulfate attack

Internal sulfate attack caused by excessive sulfate in clinker and cement, or possibly in other concrete making materials, is uncommon. However, because of the documented increased amounts of sulfates, primarily alkali sulfates, in modern cements, all necessary precautions should be taken. Such precautions include selection of quality concrete ingredients, proper proportioning of the concrete mixture, adherence to best curing practices to give dense and

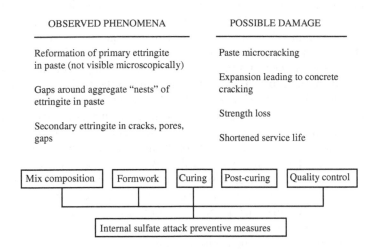

*Figure 4.16* Summary of internal sulfate attack-related issues.

impermeable concrete, and proper maintenance – all relative to the conditions of the concrete use. This is specifically recommended for concrete that might be heat cured or exposed to changing temperature and humidity.

The mechanism of so-called *delayed* ettringite formation, believed to be in some instances responsible for deterioration of heat-treated concrete structures and products, is still considered to be a controversial issue. Continuous publication of large amounts of contradictory papers is a reflection of this fact. However, the present authors believe that most basic aspects of the mechanism are now clear and that the available knowledge, once properly distributed and adopted, will become an adequate remedy for protection of concrete against heat-induced expansion and cracking. It is expected that the remaining open questions will be answered in not too distant future, but such new data, we believe, will only confirm what is known today.

Figure 4.16 reemphasizes schematically the most important aspects of heat-induced sulfate attack (DEF).

## 4.9  EXTERNAL SULFATE ATTACK

In external sulfate attack, sulfates from an external source enter the hardened concrete, causing its impairment. The deleterious action may include an excessive expansion, crack formation, loss of strength or surface spalling and delamination.

The most common external source of sulfates is ground water. Here, the sulfate anions are most commonly combined with alkali or calcium cations. In some waters the amount of present $Mg^{2+}$ ions may also be significant.

Others may contain sulfates even in the form of ammonium sulfate that entered the water after it had been used as a fertilizer. Only seldom ground water may contain also free sulfuric acid. The damage to concrete that may be expected, will depend not only on the concentration of these species in the water, but also on the water permeability of the soil adjoining the concrete structure and the amount of water that gets in contact with the concrete surface (Taylor 1997; Ferraris *et al.* 1997; Marchand and Skalny 1999).

In addition to ground water, water of rivers and lakes may also serve as source of sulfates that may cause external sulfate attack. The chemical composition of these waters varies within the range commonly seen also in ground waters, in most instances. Here, the expected damage to concrete will depend, in addition to the amount of sulfates dissolved in the water, greatly on whether the water is stagnant or flowing and, in the latter case, on the velocity of the flow.

Industrial waste waters have to be considered to be a special case. Their composition may vary greatly, depending on the kind of industry involved, and along with it also varies their harmfulness.

A special consideration has to be given to sea water, as it contains a variety of constituents, including sulfate ions, and its overall content of dissolved species is significantly higher than is common in ground, lake or river waters.

In some geographical regions rain water may contain distinct amounts of sulfuric acid. Such situation exists especially in the vicinity of power plants that use solid fuels as their energy source and are not equipped with desulfurization units.

Waters involved in an external sulfate attack contain constituents that may be detrimental to concrete besides excessive amounts of sulfates. In such cases the detrimental action of the sulfates, combines with that of the other constituent and the overall effect is usually, though not always, more damaging.

In an external chemical attack on an adequately compacted concrete body, the sulfate–cement paste interaction is initiated at the surface of the material, moving gradually inwards, as the process continues. If the corrosive action consists of more than one step, the region in which the first step takes place is located farthest away from the original surface, followed by the region in which step 2 takes place, etc. The process is accompanied by the migration of the corrosive agent inwards and, in some instances, also with the migration of some of the products of these reactions in the opposite direction. The rate of migration will depend on the paste porosity and its permeability to liquids that, in some instances, may also be affected by filling of the existing pores with reaction products (Gospodinov *et al.* 1999). It was reported that in a cement paste with a low w/cm ratio, immersed in a 0.25 molar $Na_2SO_4$ solution, the reaction front penetrated about 0.5–1.0 mm deep after six months (Gollop and Taylor 1992). In our own experiments the following average corrosion depths were found in a paste made from a sulfate-resistant cement immersed in a 0.5 molar $Na_2SO_4$ solution: 3 d.: 0.26 mm; 28 d: 0.60 mm; 106 d: 1.1 mm (Werner *et al.* 2000).

If the concrete undergoing chemical attack is insufficiently compacted, the corrosive agent may penetrate and cause damage in greater depth and the damage to concrete may be more extensive.

In most instances an external chemical attack involves only the hardened cement paste constituting the concrete. However, in rare situations, the aggregate present may also become subject of a detrimental chemical inter-action.

In the following sub-sections the chemical attack caused by different sul-fates is discussed. The discussion is limited to calcium silicates-based cements, such as all types of Portland cements, composite cements, belite cements, etc.

## 4.9.1  $Na_2SO_4$ and $K_2SO_4$

In sulfate attack caused by alkali sulfates, sulfate ions from waters containing sodium or potassium sulfate migrate into the concrete and react with mono-sulfate that had been formed in the hydration process:

$$2SO_4^{2-}+Ca_4Al_2(OH)_{12}\cdot SO_4\cdot 6H_2O+2Ca^{2+} \rightarrow Ca_6Al_2(OH)_{12}(SO_4)_3\cdot 26H_2O$$
$$\text{monosulfate} \qquad\qquad\qquad \text{ettringite}$$

More seldom, other aluminum-bearing species, such as residual of unreacted tricalcium aluminate, may serve as the source of $Al^{3+}$ ions.

The $Ca^{2+}$ ions needed for this reaction are supplied by the dissolution of the present portlandite

$$Ca(OH)_2 \rightarrow Ca^{2+} + 2OH^-$$

or – after the consumption of this phase – they may also originate from the decomposition of the existing C-S-H phase. In this process the $CaO/SiO_2$ ratio within this phase declines and this may be associated with a gradual loss of its bonding properties. The latter reaction, however, may become signific-ant only at very high alkali sulfate concentration in the water being in con-tact with the concrete.

If the amount of $Al^{3+}$ in the zone undergoing interaction with $SO_4^{2-}$ ions becomes consumed, yet still additional amounts of sulfate ions are available, gypsum rather than ettringite starts to be formed:

$$SO_4^{2-} + Ca^{2+} + 2H_2O \rightarrow CaSO_2\cdot 2H_2O$$

Thus, in concrete undergoing sulfate attack, this phase may be found closer to the concrete surface than ettringite. In some instances, however, immediately below the concrete surface the decalcification may reach a degree at which even calcium sulfate cannot be formed anymore, due to the unavailability of the needed $Ca^{2+}$ ions.

Ultimately, the following zones may be distinguished in a Portland cement paste undergoing alkali sulfate attack:

- the original paste not involved in the corrosion process;
- a zone in which ettringite had been formed in a reaction with monosulfate; the amount of calcium hydroxide is reduced;
- a zone containing gypsum; calcium hydroxide is absent, the C-S-H phase is partially decalcified (formation of horizontal cracks preferentially in this region);
- a zone containing the C-S-H phase with a significantly reduced C/S ratio as its main constituent. Limited amounts of sulfates in adsorbed form may also be present.

As to the alkali ions that had been combined with the sulfate ions in the corrosive solution, they also tend to migrate into the pore system of the cement paste, thus increasing the alkali concentration and the pH value of the pore solution. This, under unfavorable conditions, may cause a concomitant alkali–silica reaction in the concrete (Pettifer and Nixon, 1980). A limited migration of $OH^-$-ions outwards may also take place. This, under laboratory conditions, causes an increase of the pH of the sulfate containing solution in which test specimens are immersed. All the reactions taking place in Portland cement attack by alkali sulfates are summarized in Figure 4.17.

The initial manifestation of the action of alkali sulfates on concrete that was made with Portland cement is a temporary increase of strength within the affected region, brought about by filling the existing pores with ettringite

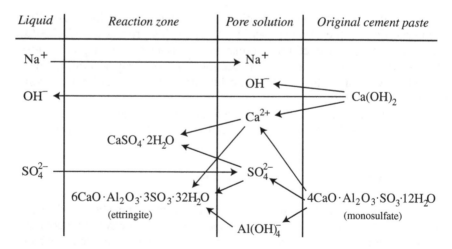

*Figure 4.17* Reactions taking place between Portland cement components and sodium sulfate solution.

(Brown 1981; Brown and Taylor 1999 and others). However, as the formation of this phase continues and the available pore space loses its capacity to accommodate additional amounts of ettringite, potentially damaging expansion forces start to be generated within the material.

The main manifestation of the damaging action is surface scaling, brought about by the expansion associated with the formation of ettringite in the underlying region. It was also suggested that crystallization pressure generated when gypsum precipitates in the pores of the hardened cement paste may play an important role in this process as well, especially in cements low in $Al_2O_3$ and at high sulfate concentrations (Gollop and Taylor 1992–1995).

Parallel to the damage caused by expansive reactions, in regions adjacent to the surface exposed to sulfates, the concrete loses strength due to the formation of microcracks and loss of bonding capacity associated with decalcification of the C-S-H phase (Gollop and Taylor 1992–1995; Mehta 1992 and others). In concrete made with Portland cement, the latter effect is usually less critical than that caused by expansive reactions.

At very high sodium concentrations of the pore solution, and especially at elevated temperatures, a separate form of sulfate attack, associated with the formation of the so-called *U-phase*, may occur. Conditions for the formation of this phase are present if the concrete body is percolated with $Na_2SO_4$-rich water, which evaporates on the opposite side of the body, or when large amounts of a concreted admixture with a high $Na^+$-content are added to the fresh concrete mix. The formation of the U-phase was also observed in cement-stabilized wastes that contained high amounts of $Na_2SO_4$ (Li *et al.* 1996a,b).

The U-phase [$4CaO \cdot 0.9Al_2O_3 \cdot 1.1SO_3 \cdot 0.5Na_2O \cdot 16H_2O$] is a Na-substituted AFm phase whose structural formula may be written as (Moranville and Li 1999):

$$\left\{Ca_4Al_{2(1-x)}(OH)_{12}[SO_4]\right\}^{6x-} \left\{yNa_2SO_4 \cdot 6x\text{Ns.Aq}\right\}^{6x+}$$

or as

$$4CaO \cdot (1-x)Al_2O_3 \cdot (1-y)\,SO_3 \cdot (3x+y)\,Na_2O$$

Thermodynamic studies (Damidot and Glasser 1993) revealed the stability of the U-phase and loss of stability of the ettringite phase in the system $CaO–Al_2O_3–CaSO_4–H_2O$ in the presence of $Na_2O$. In laboratory experiments the formation of the U-phase could be demonstrated in concrete samples that were allowed to hydrate in a 1 M NaOH solution both at 20 °C and 80 °C (Moranville and Li 1999).

The U-phase may induce deterioration of concrete through two mechanisms: First a deterioration may result if the U-phase is formed in excessive amounts, and second, if it transforms to ettringite (Li *et al.* 1996;

Moranville and Li 1999). A conversion to ettringite may take place if the alkalinity of the cement paste containing the U-phase becomes reduced. This may happen if concrete that contains this phase is leached or percolated with water.

If water evaporates from the surface of concrete whose pore solution contains high amounts of alkali sulfates a crystallization of these salts may take place in regions close to the surface. Under these conditions the crystallization pressure generated in the process may act disruptively on the material (Pettifer and Nixon 1980). A particularly unfavorable situation may occur if sodium sulfate present in the pore solution is allowed to crystallize at temperatures above about 32 °C and the formed anhydrous solid gets subsequently in contact with water (or additional amounts of the sulfate solution) at lower temperatures. Under these conditions re-crystallization of the anhydrous sodium sulfate to its hydrous form may takes place

$$Na_2SO_4 + 10H_2O \rightarrow Na_2SO_4 \cdot 10H_2O$$
$$\text{thenardite} \qquad\qquad \text{mirabilite}$$

and this reaction is associated with an increase of the solid volume by 315.4%. The disruptive action on concrete may be particularly enhanced if the existing temperature fluctuates around the above critical temperature, as under these conditions the conversion from thenardite to mirabilite and back, may repeat itself. See also Rodrigez-Navarro *et al.* (2000).

The resistance of Portland cement to alkali sulfate attack may vary, depending on its composition. Generally, the resistance improves with a declining content of $Al_2O_3$ in clinker, in particular of that, existing in form of tricalcium aluminate. Under these conditions, less monosulfate (that may be subsequently converted to ettringite upon contact with water that contains alkali sulfates) will be present in the hydrated cement paste. At the same time, the loss of cohesion and strength, due to decalcification/decomposition of the C-S-H phase, will also be reduced as less calcium oxide will be consumed for ettringite or gypsum formation (Gollop and Taylor 1992–1996). As an overall result, the expansion forces tending to disrupt the paste will be reduced and the capacity of the paste to withstand these forces will be impeded to a lesser degree.

The alkali sulfate resistance of the cement may also be improved by increasing its gypsum content, while still staying within the acceptable $SO_3$ range (Gollop and Taylor 1992–1996). By this measure the amount of $Al_2O_3$ which will remain in the mature cement paste in form of ettringite and will not convert to monosulfate in the course of hydration becomes increased. Consequently, the amount of the latter phase, available for a later conversion to ettringite if the concrete gets into contact with a sulfate-containing water, is even further reduced.

Alkali sulfate attack on Portland-slag cement proceeds in a broadly similar way as in plain Portland cement, however, there exist some distinct differences. Out of the various forms in which $Al_2O_3$ may be present in the hydrated cement paste, only monosulfate [$C_4A\bar{S}H_{12}$] may serve as a significant direct $Al^{3+}$ source for ettringite formation in a reaction with sulfate ions entering the concrete body from outside. Neither $Al^{3+}$ incorporated in the C-S-H phase, nor that present in hydrotalcite is available for such a reaction, just as $Al^{3+}$ present in ettringite formed independently of sulfate attack (Gollop and Taylor 1992–1996). Out of the $Al^{3+}$ ions originally present in the glass phase of the granulated slag, only those that became incorporated into the monosulfate phase after the slag had participated in the hydration process, may be converted to ettringite. As in the hydration of the slag usually less monosulfate is formed than in the hydration of the same amounts of clinker, the quantity of newly-formed ettringite, and thus the extent of expansion and scaling, become reduced, with increasing slag content in cement. At the same time, at equal amounts of slag in cement the amount of formed ettringite, and the observed damage to concrete, will increase with increasing $Al^{3+}$ content in the original slag (Gollop and Taylor 1992–1996).

As calcium hydroxide is also needed for the formation of ettringite, consumption of free calcium hydroxide and a subsequent decalcification of the C-S-H phase also take place in alkali sulfate attack. The degree of decalcification may be quite extensive in cements with high slag contents, in spite of relatively low amounts of ettringite formed, as the amount of formed C-S-H, and especially that of free calcium hydroxide, is also reduced in the hydration of cements containing slag. For the same reasons, the amount of gypsum in the corrosion zone of Portland-slag cement that underwent alkali sulfate attack tends to be also low.

Generally, concrete damage produced in alkali-sulfate attack tends to be reduced if slag-Portland cement, rather than ordinary Portland cement is employed. At the same time weakening of the C-S-H matrix is relatively more important than in the latter binder. Thus, disintegration of the matrix and its softening are in the foreground of damage of concrete made with slag-Portland cement, whereas cracking, scaling and expansion prevail if ordinary Portland cement was used.

In cements that contain natural or artificial pozzolanas the response to sulfate attack differs little from that seen in ordinary Portland cement. The main differences consist in a reduced amount of $C_3A$ present in cement and thus reduced amounts of formed ettringite.

### 4.9.2 CaSO$_4$

In sulfate-bearing waters in which the charge of the sulfate ion is balanced exclusively or predominantly by $Ca^{2+}$ ions, the sulfate concentration is limited by the solubility of calcium sulfate in water, which amounts to 15.2 mmol/l $CaSO_4$ (= 1.46 g $SO_4^{2-}$/l) at 20 °C. Waters of this type may occur in regions where anhydrite or gypsum is present in the nature.

In concrete exposed to water that contains calcium sulfate, the existing monosulfate phase becomes converted to ettringite in a surface-close region whose thickness increases with time:

$$4CaO, Al_2O_3 \cdot SO_3 \cdot 12H_2O + 2Ca^{2+} + SO_4^{2-} + 24H_2O$$

monosulfate $\rightarrow 6CaO \cdot Al_2O_3 \cdot SO_3 \cdot 32H_2O$

ettringite

As the whole amount of $Ca^{2+}$ needed for monosulfate to ettringite conversion is supplied by calcium sulfate, no additional $Ca^{2+}$ originating from calcium hydroxide or the C-S-H phase is needed. Thus, unlike with alkali sulfates, no decalcification of the C-S-H phase takes place in calcium sulfate attack and the integrity of this phase remains preserved.

The initial manifestation of the interaction between a Portland cement paste and calcium sulfate solution is an increase of strength, as the pores of the paste become filled with newly-formed ettringite, without producing any significant stresses. However, if the formation of ettringite continues even after the capacity of the pore space to accommodate additional amounts of this phase had been exhausted, expansive stresses develop, which may cause an expansion and eventually cracking within the material. If, after filling the pore space, an evaporation of water from the surface of the paste takes place, a crystallization of gypsum within the pore space may also occur, which will affect adversely the volume stability and cohesion of the material by generating crystallization pressure.

### 4.9.3    MgSO$_4$

The primary step in the interaction between hydrated cement paste and a magnesium sulfate solution is a reaction of the sulfate with calcium hydroxide of the paste, yielding magnesium hydroxide [brucite, $Mg(OH)_2$] and calcium sulfate in form of gypsum [$CaSO_4 \cdot 2H_2O$]:

$$Mg^{2+} + SO_4^{2-} + Ca(OH)_2 + 2H_2O \rightarrow Mg(OH)_2 + CaSO_4 \cdot 2H_2O$$

Out of these, magnesium hydroxide is practically insoluble, while calcium sulfate possesses a limited, but distinct solubility.

Both parallel and following the reaction with calcium hydroxide, a gradual decomposition of the C-S-H phase also takes place. Ultimately, this phase converts to an amorphous hydrous silica [$SiO_2 \cdot aq$] or to a magnesium silicate hydrate phase that was identified as a poorly crystalline serpentine [$M_3S_2H_2$], or both (Bonen and Cohen 1992; Gollop and Taylor 1992–1996; Brown and Taylor 1999). At the same time additional amounts of gypsum, brucite and magnesium silicate hydrate are also formed.

$$x\text{Mg}^{2+} + x\text{SO}_4^{2-} + x\text{CaO} \cdot \text{SiO}_2 \cdot \text{aq} + 3x\text{H}_2\text{O} \rightarrow x\text{CaSO}_4 \cdot 2\text{H}_2\text{O} + \\ x\text{Mg(OH)}_2 + \text{SiO}_2 \cdot \text{aq}$$

$$2x\text{Mg}^{2+} + 2x\text{SO}_4^{2-} + 2[x\text{CaO} \cdot \text{SiO}_2 \cdot \text{aq}] + y\,\text{H}_2\text{O} \rightarrow 3\,\text{MgO} \cdot 2\text{SiO}_2 \cdot 2\text{H}_2\text{O} + \\ 2x[\text{CaSO}_4 \cdot 2\text{H}_2\text{O}] + (2x-3)\,\text{Mg(OH)}_2$$

As a consequence of these reactions, in a partially corroded concrete the C/S ratio of the existing C-S-H phase is not constant; it is highest in the central core and declines towards the sample surface, attaining zero value in regions in which the corrosion process is completed.

The degradation of the C-S-H phase in the presence of magnesium sulfate is significantly faster and more thorough than with other sulfate compounds. The reason is the extremely low water solubility of magnesium hydroxide and the low pH of the solution in equilibrium with this phase. Only about 0.01 g/l of magnesium hydroxide dissolves in water at ambient temperature, and the saturated solution has a pH of 10.5, which is too low to maintain the stability of the C-S-H phase. Thus, after all the free calcium hydroxide had been consumed and the pH of the liquid phase had dropped under the stability range of C-S-H, this phase liberates calcium hydroxide in an effort to establish its equilibrium pH. This calcium hydroxide, however, is immediately also converted to magnesium hydroxide and calcium sulfate, as long as free magnesium sulfate is available, thus leading eventually to a complete degradation of the C-S-H phase.

The regions in which the chemical attack by magnesium sulfate resulted in a significant drop of the pH value, the calcium aluminate sulfate phases are also decomposed and the released $\text{Ca}^{2+}$ and $\text{SO}_4^{2-}$ ions are precipitated as gypsum (Brown and Taylor 1999; Gollop and Taylor 1992–1996). Only in regions that are sufficiently distant from those in which the MgO containing phases are precipitated, the pH value remains high enough to allow the formation of ettringite in a reaction of the sulfate ions migrating inwards, with monosulfate and portlandite. The amount of formed ettringite is, however, low, as the hardened cement paste tends to disintegrate, owing to the degradation of the C-S-H phase, before significant amounts of ettringite may be formed. All the chemical reactions taking place in magnesium sulfate attach are schematically shown in Figure 4.18.

The action of magnesium sulfate is accompanied by a migration of hydroxide ions towards the surface to produce insoluble brucite and by a migration of sulfate ions inwards to form gypsum, and in deeper regions and in smaller amounts also ettringite. The calcium ions needed for the formation of these phases are supplied primarily by the decomposition of calcium hydroxide and, especially after exhaustion of this phase, also by the decalcification of the C-S-H phase. Eventually, a double surface layer is formed consisting of an external layer of brucite, followed by a layer of gypsum.

In a recent publication, Clark and Brown (2000) discuss the reaction mechanisms of $\text{C}_3\text{A}$–magnesium sulfate interactions in different $\text{MgSO}_4$

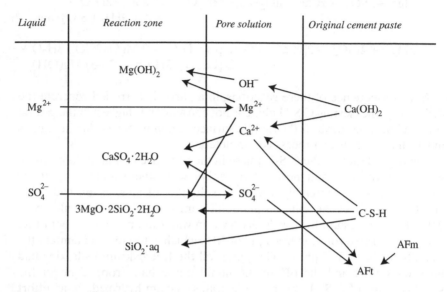

*Figure 4.18* Reactions taking place between Portland cement components and magnesium sulfate solution.

solutions. $C_3A$ hydration in 3.0 M $MgSO_4$ resulted in ettringite and gypsum formation.

$$C_3A + MgSO_4 \rightarrow gypsum + C_3AH_6$$
$$\rightarrow AFt + gypsum\ (residual) + (Mg\text{-}rich)\ amorphous\ phase(s).$$

In general, the reaction rate increased with temperature, indicating an interfacial controlled process. $C_3AH_6$ has not been observed as a final hydration product.

Monosulfate and $C_3AH_6$ were produced when hydration was carried out in 0.5 M and 1.0 M $MgSO_4$ solutions.

$$C_3A + MgSO_4 \rightarrow AFm + C_3AH_6 + amorphous\ phase(s)$$

This is interesting in that monosulfate and $C_3AH_6$ are not expected to coexist at equilibrium, and their coexistence is assumed to be due to the presence of magnesium cations. No magnesium-rich phases were detected, regardless of temperature or magnesium concentration by XRD, indicating they are amorphous.

Because of the simultaneous significant decomposition of the C-S-H phase that accompanies the formation of gypsum and ettringite, the overall corrosive action of magnesium sulfate is greater than that of alkali or calcium sulfates (at equal $SO_4^{2-}$ concentrations). Unlike the attack of alkali sulfates, that by magnesium sulfate is characterized mainly by a loss of strength and disintegration of the concrete under attack, rather than by an expansion and scaling.

### 4.9.4 $H_2SO_4$

There are several sources of sulfuric acid that may attack concrete structures:

- Free sulfuric acid may be present in some ground waters. Here it may be formed in the oxidation of sulfides, particularly of pyrite, present in some soils, by oxygen of air, under conditions of weathering (Casanova *et al.* 1996; Hobbs and Taylor 2000; Oberholster *et al.* 1983 and others)

$$2FeS_2 + 7O_2 + 2H_2O \rightarrow 2FeSO_4 + 2H_2SO_4$$

or

$$4FeS_2 + 15O_2 + 2H_2O \rightarrow 2Fe_2(SO_4)_3 + 2H_2SO_4$$

- Sulfuric acid may be present in some industrial waste-waters, after being produced in an industrial process.
- Sulfuric acid may be formed in sewers. The formation of this acid starts by the action of sulfate-reducing bacteria, e.g. *desulfovibrio*, that reduce the sulfur compounds present in sewage to $H_2S$, which escapes into the sewer atmosphere. Here $H_2S$ can react with oxygen to elementary sulfur that is deposited on the sewer wall, where it becomes available as a substrate for oxidising bacteria, *Thiobacilli*, that convert it to sulfuric acid (see also Section 9.12).
- Sulfuric acid may be present in rain water after being formed by oxidation of $SO_2$ that had been produced in the burning of fossil fuels. By the presence of this acid the pH of the rain water may be lowered and, under extreme conditions, may reach the value of pH = 4.

Unlike other types of sulfate attack, a chemical corrosion by sulfuric acid is not a pure sulfate attack, but a combined acid-sulfate attack, in which the acid component plays a significant role in the overall corrosion mechanism. More about acidic corrosion can be found in the review paper by Allahverdi and Škvara (2000).

In an attack caused by sulfuric acid, as a first step calcium sulfate (gypsum) is formed in a reaction of the acid with calcium hydroxide and the C-S-H phase, the latter being converted ultimately to amorphous hydrous silica.

$$Ca(OH)_2 + H_2SO_4 \rightarrow CaSO_4 \cdot 2H_2O$$

$$xCaO \cdot SiO_2 \cdot aq + xH_2SO_4 + xH_2O \rightarrow xCaSO_4 \cdot 2H_2O + SiO_2 \cdot aq$$

Under the conditions of low pH, brought about by the presence of the acid, the calcium sulfoaluminate hydrate phases AFm and AFt present in the hydrated paste loose their stability and convert to gypsum and aluminum sulfate. Simultaneously, however, a limited formation of ettringite may take place in deeper sections of the concrete body in which a high enough pH value is still maintained, as long as the calcium sulfate that had been formed first, can migrate into these regions.

The chemical reactions taking place result in a profound degradation of the hydrated cement paste, associated with a loss of strength. If the concrete surface undergoing corrosion by sulfuric acid is simultaneously exposed to flowing water, the products of degradation, particularly alkali sulfates, aluminum sulfate and in a lesser extent also calcium sulfate, are carried away to a significant degree.

In concrete mixes in which an acid-soluble aggregate has been employed, this aggregate may undergo decomposition along with the cement paste, in a sulfuric acid attack. Susceptible to this type of attack are all carbonate-based rocks, such as limestone [$CaCO_3$], dolomite [$CaMg(CO_2)_2$], magnesite [$MgCO_3$] etc. Generally, an attack by free sulfuric acid is more severe than any with a neutral sulfate solution.

### 4.9.5   $(NH_4)_2SO_4$

Just as the attack with sulfuric acid, the corrosion of concrete by ammonium sulfate is also a combined sulfate-acid attack in its nature (Guerrero *et al.* 2000; Miletic and Ilic 1997). The first step in the process is a reaction of the sulfate with calcium hydroxide of the paste, resulting in the formation of calcium sulfate (in form of gypsum) and the release of gaseous ammonia:

$$Ca(OH)_2 + (NH_4)_2SO_4 \rightarrow CaSO_4 \cdot 2H_2O + 2NH_3$$

This reaction takes place until calcium hydroxide becomes completely consumed, resulting in a decrease of the existing pH.

Parallel and subsequently to the reaction with calcium hydroxide, ammonium sulfate reacts also with the C-S-H phase, causing initially a decrease of the C/S ratio and eventually a conversion of this phase to amorphous hydrous silica:

$$xCaO \cdot SiO_2 \cdot aq + x(NH_4)_2SO_4 + xH_2O \rightarrow$$
$$SiO_2 \cdot aq + xCaSO_4 \cdot 2H_2O + 2xNH_3$$

This reaction, associated with a strength loss of the material (Miletic and Ilic 1997), is promoted by the lowering of the pH of the pore solution beyond the value needed to maintain the C-S-H phase stable, which results from the preceding disappearance of calcium hydroxide.

Due to the lowering of the pH, the present calcium sulfoaluminate hydrate phases AFm and AFt become unstable and convert to gypsum and aluminum sulfate. However, in deeper regions of the cement paste, in which a high enough pH is still maintained a formation of some ettringite, in a reaction between the primary formed gypsum and monosulfate, may still take place.

Generally, the degradation of concrete by ammonium sulfate is more severe than that by alkali sulfates, but less severe than the corrosion caused by free sulfuric acid (Miletic and Ilic 1997).

### 4.9.6   Sulfate attack in the presence of $SiO_2$ and $CO_2$

In the presence of an adequate supply of sulfate and carbonate ions, and at low enough temperature and pH levels above 10.5, *thaumasite* $[Ca_3[Si(OH)_6] \cdot CO_3 \cdot SO_4 \cdot 12H_2O$, abbr. $3CaO \cdot SiO_2 \cdot CO_3 \cdot SO_3 \cdot 15H_2O$ or $C_3S\bar{C}\bar{S}H_{15}]$ may be formed in a Portland cement based concrete:

$$3Ca^{2+} + SiO_3^{2-} + CO_3^{2-} + SO_4^{2-} + 15H_2O \rightarrow 3CaO \cdot SiO_2 \cdot CO_2 \cdot SO_3 \cdot 15H_2O$$
$$\text{thaumasite}$$

The occurrence of thaumasite sulfate attack (TSA) in the field has rarely been reported in the literature. However, a growing series of cases have been discovered in recent years in the United Kingdom. This prompted the UK Government to convene a Thaumasite Expert Group in 1998 to investigate the seriousness of the problem and to provide new interim recommendations in order to minimize the effect of TSA in new construction.

In severe cases of TSA, the hardened cement paste matrix is completely replaced by thaumasite and the once strong "glue" which held the concrete together is transformed into a white, incohesive mush.

In all the field and laboratory studies investigated so far, the sulfates needed to fuel the thaumasite form of sulfate attack have never originated from a "normal" cement paste. They can be present as an internal ingredient of a concrete but only if the cement or aggregate is contaminated with gypsum or similar sulfate-bearing material. In the majority of TSA occurrences in the field, the sulfates originated from an external source; e.g. ground water, sulfate-bearing bricks, gypsum plaster and gypsum-bearing historical mortars (TEG Report 1999). However, contact with sulfate-containing ground water remains the most common cause of the thaumasite form of sulfate attack in buried concrete (Crammond and Nixon 1993, TEG Report 1999).

The carbonate ions usually originate from the carbonate rocks constituting the concrete aggregate (Bickley *et al.* 1994; Crammond and Nixon 1993; Crammond and Halliwell 1995; Sibbick and Crammond 1999; TEG Report 1999), but calcium carbonate inter-ground with the cement, atmospheric $CO_2$, or ground water with high amounts of dissolved $CO_2/HCO_3^-$ may also serve as sources of carbonate ions. In concrete mixes containing

CaCO$_3$ in finely divided form the formation of thaumasite is particularly extensive and rapid (Crammond and Halliwell 1995). The threshold amount of carbonates in the aggregate needed to initiate thaumasite formation will depend on the fineness of the aggregate particles. It has been tentatively estimated that this amount lies somewhere between 50% and 75% for an aggregate with particles 5–20 mm and below 25% for an aggregate with particles smaller than 5 mm. (TEG Report 1999). As to the cement, it was found in laboratory experiments that its sulfate resistance decreases with increasing limestone filler content (Halliwell *et al.* 1996).

The silicate ions needed for TSA come from the C-S-H phase (Crammond and Nixon 1993; Crammond and Halliwell 1995 and others), and the degradation of this constituent of the cement paste is associated with the loss of the binding capacity of the material.

Even though pure thaumasite does not contain Al$_2$O$_3$, it appears that the presence of limited amounts of aluminate ions in the thaumasite structure is not only possible, but that a small amount of reactive alumina is a beneficial ingredient for the formation of thaumasite. It was suggested that a preceding formation of ettringite may be necessary to act as a nucleation agent for thaumasite and this would explain the need for a source of Al$_2$O$_3$.

The formation of thaumasite is favored at low temperatures and a temperature of around 5 °C is most favorable. Nevertheless, a formation of this phase at temperatures up to about 25 °C is possible. The rate of thaumasite formation drops off markedly at somewhere between 15 °C and 20 °C (Crammond and Nixon 1993; Crammond and Halliwell 1995; Halliwell *et al.* 1996; Halliwell and Crammond 1999 and others). The formation of thaumasite is usually preceded or accompanied by that of ettringite (Gaze 1997).

Thaumasite is a phase structurally similar to ettringite. In its structure Al(OH)$_6^{3-}$ ions are replaced with Si(OH)$_6^{2-}$ ions and $(3SO_4^{2-} + 2H_2O)$ with $(2CO_3^{2-} + 2SO_4^{2-})$. Nevertheless, in spite of the existing similarities, both phases may be easily distinguished by X-ray diffraction (Crammond 1985a; TEG 1999). Under field conditions, thaumasite and ettringite may form intimate mixtures and solid solutions.

The damage to concrete due to TSA is usually profound and is mainly due to the decomposition of the C-S-H phase, resulting in a loss of strength of the material. As this process does not involve the aluminate component, the use of "sulfate resistant" Portland cement, such as ASTM Type V cement, does not offer an improvement of the resistance to this type of sulfate attack (Crammond and Nixon 1993; Halliwell and Crammond 1999; Sibbick and Crammond 1999). On the other hand it was reported that concrete mixes made with blended cements containing high amounts of granulated blast furnace slag performed reasonably well, as long as the present limestone aggregate was of high quality (Halliwell and Crammond 1999; see also Case Study in Chapter 8).

### 4.9.7    Sea water

The corrosion of concrete by sea water is the result of the simultaneous action of several ions present in the water in different concentrations. The main ions present are $Na^+$, $Mg^{2+}$, $Cl^-$ and $SO_4^{2-}$, together with smaller amounts of $K^+$, $Ca^{2+}$, $HCO_3^-$ and $Br^-$. The typical concentration of $SO_4^{2-}$ in sea water is 2,700 mg/l.

The initial product of sea water attack on concrete are brucite $[Mg(OH)_2]$ and aragonite $[CaCO_3]$, formed by the action of $Mg^{2+}$ and dissolved $CO_2$ (Mehta and Haynes 1975; Mehta 1999; Thomas *et al.* 1999).

$$Mg^{2+} + Ca(OH)_2 \rightarrow Mg(OH)_2 + Ca^{2+}$$

$$CO_2 + Ca(OH)_2 \rightarrow CaCO_3$$

$$xCO_2 + xCaO \cdot SiO_2 \cdot aq \rightarrow xCaCO_3 + SiO_2 \cdot aq$$

The formation of metastable aragonite, in place of calcite, is promoted by the presence of $Mg^{2+}$ ions.

The products of the above reactions may produce a surface skin consisting of a layer of brucite about 30 μm thick overlain by a layer of aragonite. This skin may protect a well produced concrete from further attack in permanently submerged regions or may, at least, slow down the progress of further corrosion. Additional phases formed in sea water attack may include magnesium silicates, gypsum, ettringite and calcite, and in concrete mixes made with a carbonate-based aggregate also thaumasite (Thomas *et al.* 1999).

The formation of ettringite in sea water attack typically does not lead to expansion and cracking of the concrete and it is believed that the formation of this phase is non-expanding in the presence of excessive amounts of chloride ions (Mehta 1999; Thomas *et al.* 1999). The main deleterious effects results from the degradation of the C-S-H phase and its ultimate conversion to magnesium silicate (Mehta 1991, 1999; Thomas *et al.* 1999).

$$xCaO \cdot SiO_2 \cdot aq + 4Mg^{2+} + 4SO_4^{2-} + (4-x)\,Ca(OH)_2\,nH_2O \rightarrow$$
$$4MgO \cdot SiO_2 \cdot 8H_2O + 4CaSO_4 \cdot 2H_2O$$

This conversion is ultimately responsible for a softening of the cement paste matrix and a gradual loss of strength.

In and above the tidal zone, a repeated water uptake and evaporation may result in salt crystallization, thus enhancing the deterioration of the concrete taking place. Still additional damage may by caused by mechanical erosion due to waves, solid debris or ice.

In steel reinforced concrete a significant corrosion of the reinforcement may be induced by chloride ions that migrate into the concrete body.

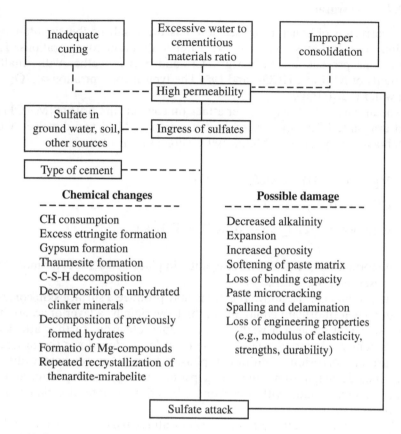

*Figure 4.19* Summary of external sulfate attack-related issues.

### 4.9.8   Concluding comments on *external* sulfate attack

Similar to *internal* sulfate attack, there are not too many reported cases of *external* sulfate attack either. Those that were reported in the literature are mostly cases where some of the basic principles of proper concrete practice were violated. The primary violations relate to excessive w/cm and inadequate or improper curing, enabling external sulfate ions to penetrate into porous concrete through its open pore structure. A brief summary of issues related to external sulfate attack is given in Figure 4.19.

### 4.10   SULFATE RESISTANT CEMENTS – MECHANISM OF ACTION

The capacity of different types of cement to sustain sulfate attack varies in a wide range. These differences are the result of dissimilarities in hydration chemistry but also in the permeability of the formed hardened cement paste. Basically, one can divide cements into three categories:

- cements sensitive to sulfate attack;
- cements with an increased resistance against sulfate attack. These cements resist sulfate attack better than cements of the first category, but may fail at too high sulfate concentrations, or under a prolonged action of sulfates, or if sulfate ions are combined with $Mg^{2+}$ cations.
- cements resistant to sulfates, which do not show any signs of deterioration even if permanently exposed to sulfates in high concentrations.

The sulfate resistance of *ordinary* Portland cement is rather limited, mainly due to the presence of significant amounts of tricalcium aluminate. In the course of hydration this phase yields first ettringite which – after the added calcium sulfate had been consumed – converts completely or to a significant degree to monosulfate. If later, in the course of service, the hardened concrete is exposed to sulfates from an external source, the present monosulfate converts back to ettringite and this reaction is associated with expansion resulting in scaling, cracking and loss of cohesion.

In *sulfate resistant* Portland cement the $Al_2O_3$ content of the clinker is reduced and this oxide is bound predominantly or exclusively within the ferrite phase. The amount of tricalcium aluminate is decreased or this phase may be even absent. Because of these measures, the amount of ettringite (AFt-phase) formed in the hydration of this cement is significantly reduced and – as the ferrite phase is the main source of $Al^{3+}$ – this phase contains significant amounts of $Fe^{3+}$, replacing $Al^{3+}$ in its crystalline lattice. As this form of AFt, has a reduced tendency to convert to monosulfate (Chen and Odler 1992; Lota *et al.* 1995), most of it remains preserved even in mature cement pastes. Consequently, the amount of monosulfate available for a reaction with sulfate ions entering from an external source is significantly reduced.

As the hydration of the ferrite phase is rather slow, distinct amounts of it may be present even in mature pastes. Upon contact with sulfate ions from an external source this fraction of the ferrite phase may react to yield ettringite. This reaction is associated with an expansion, however the expansiveness of this, iron-doped, AFt phase is rather small. From the above it is obvious that even "sulfate resistant" Portland cements may expand under particularly unfavorable conditions. However, if an expansion occurs, it progresses significantly more slowly and is usually less severe. The main reason for an improved sulfate resistance of "sulfate resistant Portland cement" appears to be the reduced amount of monosulfate present in the mature paste (Eglinton 1998; Lota *et al.* 1995; Odler and Gasser 1988). Also important may be the reduced expansivity of the $Fe^{3+}$-doped AFt, produced from the ferrite phase (Odler and Gasser 1988; Odler and Collan-Subauste 1999; Regourd *et al.* 1980). Finally, the reduced rate at which AFt is formed if the ferrite phase serves as the source of $Al^{3+}/Fe^{3+}$ may also play a role. It must also be stressed that the improved performance of "sulfate resistant Portland cement" applies only to its reduced expansiveness. It does not apply to its ability to resist degradation of the C-S-H phase, a damage mechanism typical for magnesium sulfate attack.

*Fly ash – Portland Cements*, in which up to about 30% of clinker is replaced by fly ash, resist sulfate attack better than ordinary Portland cement (Djuric *et al.* 1996; Giergiczny 1997; Irassar and Ans-Batic 1989; Krizan and Zivanovic 1997; Mangat and Khatib 1992, 1995; Miletic and Ilic 1997; Soroushian and Alhozami 1992 and others). Different fly ashes, however, may not be equally effective, class F ashes being more effective than ashes of class C (Soroushian and Alhozami 1992; Biricik *et al.* 2000). A similar effect as with fly ashes may be attained also by combining the clinker with natural pozzolanas (Sersale *et al.* 1997) or with silica fume (Akoz *et al.* 1999; Giergiczny 1997). The beneficial effect of pozzolanic materials on sulfate resistance results from the reduced amount of calcium hydroxide among the hydration products and the formation of additional amounts of the C-S-H phase. This brings about a lower porosity and reduced permeability of the hardened cement paste, thus limiting the capacity of the sulfate solution to penetrate into deeper regions of the concrete structure. A reduced $C_3A$ content in the cement also contributes to the improved sulfate resistance. To obtain a proper performance, however, a sufficient curing time prior to exposure to sulfate solutions is essential. Just as Portland cement, also cements with added pozzolanic additives are more vulnerable to magnesium sulfate than to other sulfate forms, but at very high ash contents (70%) the cement may perform still acceptably well (Krizan and Zivanovic 1997; Guerrero *et al.* 2000).

At sufficiently high slag contents (60% and more) *Portland–slag cement* exhibits a better resistance to sulfate attack than ordinary Portland cement, whereas the resistance is scarcely affected at low substitution levels (Giergiczny 1997; Gollop and Taylor 1992–1996; Krizan and Zivanovic 1997; Mehta 1992; Sersale *et al.* 1997 and others). This improved performance is mainly due to a reduced $C_3A$ content in the cement, brought about by the replacement of a fraction of the clinker with granulated blast furnace slag. This way the amount of monosulfate in the hardened paste, which is the only $Al^{3+}$ form readily available for ettringite formation, becomes reduced. As to the $Al^{3+}$ bound within the slag, most of it becomes incorporated into the C-S-H phase and the hydrotalcite-type phase in the course of hydration and is not available for ettringite formation, just as ettringite already formed independently of sulfate attack (Gollop and Taylor 1992–1996). Only the rest of $Al^{3+}$ originally present in the slag converts to monosulfate, though at a reduced rate, and may ultimately also be available for a reaction with sulfate ions originated from an external source. Consequently, even at a constant replacement level, the susceptibility to sulfate attack of slag – Portland cements will vary and will increase with increasing $Al_2O_3$ content in the slag (Mehta 1992;

Gollop and Taylor 1992–1996). An improved sulfate resistance of slag – Portland cement may be observed only in alkali or calcium sulfate attack, where the main mode of the deleterious action is the formation of ettringite. Contrary to that, in magnesium sulfate attack, in which a degradation of the C-S-H phase stands in the forefront, this type of cement performs poorly and cannot be recommended.

The high sulfate resistance to sulfate attack (except that by magnesium sulfate) of *supersulfated cement* is due to the following facts: In a completely hydrated paste virtually all $Al_2O_3$ is present in form of ettringite or is incorporated within the C-S-H phase and thus is not available for a reaction with sulfate ions. Under field conditions, however, supersulfated cement pastes tend to be incompletely hydrated even after years of hydration. Even under these conditions, they exhibit superb sulfate resistance due to a very low free calcium hydroxide content, which prevents a significant ettringite formation in the presence of sulfates.

Constructions made with *calcium aluminate cement* perform rather well if exposed to sulfate solutions, especially if made with a high cement content and low water–cement ratio. This good performance is attributed to a surface densification of the hardened material, resulting in a very low permeability of the surface layer and/or to the absence of calcium hydroxide in the system (Scrivener and Capmas 1998). Unlike Portland cement and related binders, magnesium sulfate solutions are less aggressive to calcium aluminate cement based concrete than alkali sulfate solutions. This is due to the absence of the C-S-H phase in the hydrated cement paste, as this constituent is particularly susceptible to magnesium sulfate attack.

Cements completely resistant to sulfate attack include *phosphate cements*, *alkali silicate cement*, and *geopolymer cement*. Each of them possesses a completely different mechanism of setting and hardening than binders discussed above.

## 4.11 PHYSICAL SULFATE ATTACK OR SALT CRYSTALLIZATION

As briefly discussed above, the issue of *chemical* versus *physical* sulfate attack is controversial. According to Haynes *et al.* (1996), "Physical attack on concrete by salt crystallization in pores has received little attention because it seems that the problem is often misidentified as sulfate attack." Similar sentiment was also expressed by Yang *et al.* (1997) and discussed by Mehta (2000). Recently, Hime and Mather (1999) and Haynes (2000) defended the separation of *chemical* from *physical* sulfate attack, not taking into account the chemical foundations of the observed physical damage (Skalny *et al.* 2000). In line with the results of other authors, Yang *et al.* (1997) have reported that damage due to physical sulfate attack is much more pronounced when concrete is exposed to wetting–drying cycles and can be even "larger than those by chemical attack of salt." Whereas sulfate salt crystallization is potentially a serious form of concrete deterioration (see for example Figures 4.20 and 4.21), its complete separation from chemical sulfate attack is probably incorrect and the reported misidentification unsubstantiated (Young 1998; Skalny and Pierce 1999). *Physical attack* involving sulfates is *also sulfate attack* and should be categorized as such.

According to Young (1998) "physical processes don't occur without chemical changes" and sulfate salt crystallization – especially in cases where a portion of the relevant concrete structure is exposed to frequent temperature and humidity changes – "is the result of superposition of an upward flux of sulfate driven by capillary effects on regular sulfate attack."

The chemical nature of this process is explicitly stated also in ACI's Guide to Durable Concrete (ACI 1992), Section 2.3.1:

> Where concrete structures are placed on reclaimed coastal areas with the foundation below the saline ground water levels, capillary suction and evaporation may cause supersaturation and crystallization in the concrete above ground, resulting both in *chemical attack* on the cement paste (sulfate), and in aggravated corrosion of steel (chlorides).
>
> In tropical climates these combined deleterious effects may cause severe defects in concrete in the course of a very few years.

Needless to say, this statement applies not only to sea water but also to ground waters with high concentrations of sulfate, chloride and other ionic species.

Ground water and the ionic species present in it enter and penetrate concrete by one or more of the following mechanisms: adsorption, vapor diffusion, liquid assisted vapor transfer, saturated liquid flow, or ionic diffusion under saturated conditions (Hearn and Figg 2000). Whatever the transport mechanism, the sulfates present in the penetrating solution will, under appropriate chemical conditions, react with the paste components to form reaction products that may lead to changes in microstructure and paste chemistry. The rate of these reactions will depend on the local conditions (pore structure and connectivity, temperature, humidity, concentrations of ionic species in the solution, chemistry of paste components, etc.) and the time of exposure. Some of the ionic species, typically sodium and sulfate ions, may reach the surface and will lead to formation and repeated recrystallization of thenardite to/from mirabelite, leading to what is referred to as *physical sulfate attack* or *salt crystallization* (e.g. Folliard and Sandberg 1994; Haynes *et al.* 1996). An example of such penetration leading to salt crystallization is represented in Figure 4.21. Analysis of the visible efflorescing material showed presence of analytically pure $Na_2SO_4$. Such concrete can reveal clear signs of chemical and microstructural changes, including low levels of calcium hydroxide, presence of gypsum and magnesium silicate, and excessive amounts of ettringite.

*Figure 4.20* Penetration of sulfate-bearing ground water through a retaining wall exposed to atmospheric changes (wetting–drying, heating–cooling) (Photo J. Skalny).

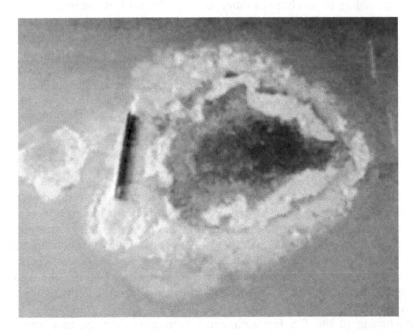

*Figure 4.21* Concrete garage slab showing damage caused by penetration of sulfates present in ground water. Efflorescing material: analytically pure sodium sulfate (Photo J. Skalny).

Similar data were presented by Haynes *et al.* (1996) – see Figure 3 of their paper – showing a garage slab made without a water barrier, and exposed to sodium sulfate in ground water. They show a salt damaged ring in the portion of the slab made with "high" (undefined!) w/c, "whereas the concrete outside the ring had a water–cement ratio of 0.62." (Incidentally, such high w/cm concrete must have been produced in violation of the Uniform Building Code (see Table 1.4). First, the concrete was obviously inhomogeneous. Secondly, if the ground water concentration was *mild* (meaning: below 150 ppm; no maximum limit on w/cm per UBC), then the concrete must have been of too high porosity (low cement content, too high w/cm, incorrect curing, or a combination of the above; or an inadequate UBC requirement) to enable efflorescence. On the other hand, if the sulfate concentration was above 150 ppm, then the concrete w/cm should have been 0.5 or below.

Binda and Baronio (1987), in their paper on the conditions under which crystallization of salts can occur, relate the possible damage to the balance between the rate of transport of the salt solution through the material (e.g. concrete) and the rate of evaporation of the water from the surface. According to these authors, if the resupply of salt solution through the material is higher than the rate of evaporation of the water from the surface, then crystallization will occur outside the material, on the surface; this process is believed not to cause mechanical damage. However, if the water evaporation rate from the surface is higher than the rate of resupply of the salt solution through the concrete, then crystallization of salts occurs within the matrix, possibly leading to mechanical damage in the form of expansion, cracking, surface erosion, and spalling. The authors do not address situations where the relative rates between evaporation and transport, and the availability of moisture on the wet side of the material (e.g. concrete), are variable due to seasonal or daily temperature and humidity changes. Neither do they discuss the possible chemical interactions between the transported salt solution (e.g. sulfates) and the matrix components.

## 4.12   BACTERIOGENIC CORROSION OF CONCRETE IN WASTE WATER NETWORKS

The internal surface of concrete sewers is frequently subject to rapid deterioration, and in extreme cases may lead to complete disintegration of the concrete product. This form of deterioration happens when the sewer atmosphere contains appreciable quantity of hydrogen sulfide and such concrete corrosion of variable degree has been observed in wastewater networks all around the world (e.g. Parker 1945). It is observable only above the water/ sewage level, with primary damage occurring just above the daily high water level.

According to the US Environmental Protection Agency, this form of concrete damage has been identified in more than twenty countries, and in more than 70% of American cities. The phenomenon consists in a gradual dissolution of the cement paste and, in concretes made with acid soluble aggregate, also in dissolution of the latter one. Where the bacteriogenic degradation has been well established, the cement paste constituents are completely broken down leaving the less acid-soluble aggregate imbedded in soft, putty-like mass. This chemical action is enhanced by the mechanical action of the flowing water that also has the capacity to carry away the products of dissolution. The result of these processes is a gradual reduction of the cross-section of the existing concrete pipes and other structural elements.

The corrosion of the concrete surface is caused by sulfuric acid, which is the endproduct in a chain of reactions brought about by the action of certain bacteria, e.g. *thiobacillus concretivorus*, present in the sewer system:

compounds of sulfur: $S \rightarrow H_2S \rightarrow SO_4^{2-} \rightarrow H_2SO_4$

The bacteriogenic corrosion of concrete is in essence a combined sulfate-acid attack. However, in waste water networks, the dissolution of the material associated with the acid action stands in the foreground due to the presence of a moving liquid phase that has the capacity to transport away the products of the acid-related dissolution reactions.

The progress of the corrosion process depends on the rate of bacteriogenic acid production, the quality – especially density or impermeability – of the cementitious binder employed, the acid solubility of the aggregate, and the amount and flow conditions of the sewage in the pipe.

Indicators of the capability of the binder to resist bacteriogenic corrosion is its stability at low pH values and its neutralization capacity, defined as the amount of acid needed to destroy/dissolve a given amount of hydrated cement. It was reported that in this respect calcium aluminate cement is more suitable than Portland cement, as its chemical stability extends to lower pH values and its neutralization capacity is higher (Scrivener 2000). The neutralization capacity of calcium aluminate cement based concrete or mortar may be increased even further, by combining this cement with a synthetic calcium aluminate based aggregate rather than with a non-reactive silicate material. In practical applications, the use of calcium aluminate cement based linings was found to be effective in protecting both iron pipes and centrifuged concrete waste water pipes from excessive bacteriogenic corrosion.

For more information see relevant references (e.g. Parker 1945; Oberholster *et al.* 1983; Hormann *et al.* 1997; Monteni *et al.* 2000 and others).

## 4.13   CONDITIONS AFFECTING SULFATE ATTACK

The conditions affecting the chemical processes of sulfate attack were discussed in more detail in Sections 3.2–3.4. It should therefore suffice to reiterate that the degree, form and kinetics of all forms of sulfate attack, and of the related damage, do depend on:

- *Quality and proportion of the concrete mixture ingredients*, including cement, supplementary materials, fine and coarse aggregate, chemical and mineral admixtures, and water.
- *Processing conditions*, including mixing, placing, curing of fresh concrete, and maintenance of hardened concrete. Special emphasis has to be placed on those processes that control the water tightness, e.g. proper consolidation and curing.
- *Environmental – atmospheric conditions* in which the concrete structure is being used. Not only are the temperature and humidity of great importance; so are their variations with time, especially in climates where capillary suction and evaporation may cause supersaturation and crystallization in the concrete above ground.

## REFERENCES

van Aardt, J.H.P. and Visser, S. (1985) "Influence of alkali on the sulfate resistance of ordinary portland cement mortars", *Cement and Concrete Research* **150**: 485–494.

ACI (1992) ACI 201.2R-92, *Guide to Durable Concrete*, ACI.

Akoz, F., Turker, F., Koral, S. and Yuzer, N. (1999) "Effects of raised temperature of sulfate solutions on the sulfate resistance of mortar with and without silica fume", *Cement and Concrete Research* **29**: 533–537.

Allahverdi, A. and Škvara, F. (2000) "Acidic corrosion of hydrated cement based materials – part 1: mechanism of the phenomenon; part 2: kinetics of the phenomenon and mathematical models", *Ceramics – Silikaty* **44**: 81–120, 121–160.

Alunno-Rosetti, V., Cioccio, G and Paolini, A.E. (1982) "Expansive properties of the mixture $C_4A\bar{S}H_{12} + 2C\bar{S}$. I. A hypothesis on expansive mechanism", *Cement and Concrete Research* **12**: 577–585.

ASTM (1995) ASTM C 150-95, Standard Specification for Cement, Table 1.

Batic, O.R., Milanesi, C.A., Maiza, P.J. and Marfil, A. (2000) "Secondary ettringite formation in concrete subjected to different curing conditions", *Cement and Concrete Research* **30**: 1407–1412.

Bentur, A. and Ish-Shalom, M. (1975) "Properties of type K expansive cement of pure components. II. Proposed mechanism of ettringite formation and expansion in unrestrained paste of pure expansive compounds", *Cement and Concrete Research* **4**: 709–721.

Bentur, A. (1976) "Effect of gypsum on hydration and strength of $C_3S$ pastes", *Journal of the American Ceramic Society* **59**: 210–213.

Bickley, J.A., Hemmings, R.A., Hooton, R.D. and Belinski, J. (1994) "Thaumasite related deterioration of concrete structures" in P.K. Mehta (ed.) *Concrete Technology: Past, Present and Future*, American Concrete Institute, Detroit, MI, SP-144, pp. 159–175.

Binda, L. and Baronio, G. (1987) "Mechanisms of masonry decay due to salt crystallization", *Durability of Building materials*, N. 4., Elsevier, Amsterdam, pp. 227–240.

Bing, T. and Cohen, M.D. (2000) "Does gypsum formation during sulfate attack on, concrete lead to expansion?", *Cement and Concrete Research* **30**: 117–124.

Biricik, H. Akoz, F., Turker, F. and Berktay, I. (2000) "Resistance to magnesium sulfate and sodium sulfate attack of mortars containing wheat straw ash", *Cement and Concrete Research* **30**: 1189–1197.

Bonen, D. and Cohen, M.D. (1992) "Magnesium sulfate attack on portland cement paste. I. Microstructural analysis. II. chemical and mineralogical analysis", *Cement and Concrete Research* **22**: 169–180 and 707–718.

Brown, P.W. (1981) "An evaluation of the sulfate resistance of cements in a controlled environment", *Cement and Concrete Research* **11**(5/6): 719–727.

Brown, P.W. (1986) "The implications of phase equilibria on hydration in the tricalcium silicate-water and tricalciumaluminate-gypsum-water systems", in *8th Int. Congress on the Chemistry of Cement*, vol. 3, Rio de Janeiro: Abla Grafica e Editora Ltda, pp. 231–238.

Brown, P.W. and Bothe, Jr. J.V. (1993) "The stability of ettringite", *Advances in Cement Research* **5**: 47–63.

Brown, P.W. and Doerr, A. (2000) "Chemical changes in concrete due to ingress of chemical species", *Cement and Concrete Research* **30**: 411–418.

Brown, P.W. and Taylor, H.F.W. (1999) "The role of ettringite in external sulfate attack", in J. Marchand and J. Skalny (eds) *Materials Science of Concrete Special Volume: Sulfate Attack Mechanisms*, The American Ceramic Society, Westerville, OH, pp. 73–98.

Casanova, I., Agullo, L. and Aguado, A. (1996) "Aggregate expansivity due to sulfide oxidation – I. Reaction system and rate model", *Cement and Concrete Research* **26**: 993–998.

Chatterji, S.K. and Jeffery, J.W. (1963) "New hypothesis on sulfate expansion", *Magazine of Concrete Research* **15**: 83–86.

Chatterji, S.K. and Jeffery, J.W. (1967) "Further evidence relating to the 'new hypothesis on sulfate expansion'", *Magazine of Concrete Research* **19**: 185–189.

Chatterji, S.K. (1968) "Mechanism of sulfate expansion of hardened cement paste", *Proceedings of the 7th International Symposium on the Chemistry of Cement*, Tokyo, **2**: 336–341.

Chatterji, S. and Thaulow, N. (1997) "Unambiguous demonstration of destructive crystal growth pressure", *Cement and Concrete Research* **27**: 811–816.

Chen, Y. and Odler, I. (1992) "The progress of Portland cement hydration: Effect of Clinker composition", *Proceedings of the 9th International Congress on the Chemistry of Cement*, New Delhi, pp. 25–30.

Clark, B.A. and Brown, P.W. (1999) "The formation of calcium sulfoaluminate hydrate compounds, part I", *Cement and Concrete Research* **29**: 1943–1948.

Clark B.A. and Brown, P.W. (2000) "The formation of ettringite from tricalcium aluminate and magnesium sulfate", *Adv. In Cement Research* **12**(4): 137–142.

Clark, B.A., Draper, E.A., Lee, R.J., Skalny, J., Ben-Bassat, M. and Bentur, A. (1992) "Electron-optical evaluation of concrete cured at elevated temperatures", in C. McInnis (ed.) *ACI SP-139-3* American Concrete Institute, pp. 41–59.

Clastres, P., Murat, M. and Bachiorini, A. (1984) "Hydration of expansive cements – Correlation between the expansion and formation of hydrates", *Cement and Concrete Research* **14**: 199–206.

Collepardi, M. (1997) "A holistic approach to concrete damage induced by delayed ettringite formation", *Proc. 5th CANMET/ACI Int. Conf. on Superplasticizers and Other Chem. Admixtures in Concrete*, ACI SP-173, American Concrete Institute, Detroit, pp. 373–396.

Collepardi, M. (1999a) "Concrete sulfate attack in a sulfate-free environment", *Proc. 2nd CANMET/ACI Int. Conf. on High-Performance Concrete*, Gramado, RS., Brazil, ACI SP-186, American Concrete Institute, Detroit, pp. 1–20.

Collepardi, M. (1999b) "Damage by delayed ettringite formation", *Concrete International* **21**: 69–74.

Crammond, N.J. (1985a) "Quantitative X-ray diffraction analysis of ettringite, thaumasite and gypsum in concretes and mortars", *Cement and Concrete Research* **15**: 431–441.

Crammond, N.J. (1985b) "Thaumasite in failed cement mortars and renders", *Cement and Concrete Research* **15**: 1039–1050.

Crammond, N.J. and Nixon, P.J. (1993) "Deterioration of concrete foundation piles as a result of thaumasite formation", *Proceedings 6th International Conference on Durability of Building Materials*, Tokyo, pp. 295–305.

Crammond, N.J. and Halliwell, M.A. (1995) "The thaumasite form of sulfate attack in concretes containing a source of carbonate ions – microstructural overview", *Proceedings 2nd Symposium on Advances in Concrete Technology*, ACI, P154-19, pp. 357–380.

Damidot, D. and Glasser, F.P. (1992) "Sulphate attack on concrete: prediction of the aft stability from phase equilibria", *International Congress on the Chemistry of Cement*, New Delhi, vol. V, pp. 316–321.

Damidot, D. and Glasser, F.P. (1993) "Thermodynamic investigation on the $CaO-Al_2O_3-CaSO_4-H_2O$ system at 25 °C and the influence of $Na_2O$", *Cement and Concrete Research* **23**: 221–238.

D'Ans, J. and Eick, H. (1953) "The system $CaO-Al_2O_3-CaSO_4-H_2O$ at 20 °C" (in German), *Zement-Kalk-Gips* **6**: 302–306.

Degenkolb, M. and Knoefel, D. (1996) "Expansive additives in slag-based Mortars", *Proceedings 18th International Conference on Cement Microscopy,* pp. 304–316.

Day, R.L. (1992) "The effect of secondary ettringite formation on the durability of concrete: a literature analysis", *PCA R&D Bulletin RD108T.*

Deng, M. and Tang, M. (1994) "Formation and expansion of ettringite crystals", *Cement and Concrete Research* **24**: 119–126.

DePuy, G.W. (1997) "Review of sulfate attack in US and Canada", Report submitted on behalf of plaintiffs in *Murphy v. First Southwest Diversified Partners*, Case #752593 (Orange County Superior Court, California), 3 November 1997, 89 pp.

Detwiler, R.J. and Powers-Couche, L.J. (1996) "Effect of sulfates in concrete on its resistance to freezing and thawing", in B. Erlin (ed.) *Ettringite: The Sometimes Host of Destruction*, ACI SP-177, pp. 219–247.

Diamond, S. (1996a) "Delayed ettringite formation – processes and problems", *Cement and Concrete Composites* **18**: 205–215.

Diamond, S. (1996b) "Delayed ettringite formation – microstructural studies of occurrences in steam cured and non-steam cured concretes", in *Proc. 18th Int. Conf. Cement Microscopy*, ICMA, p. 379.

Diamond, S. (1976) "Cement paste microstructure – an overview at several levels", in *Hydraulic Cement Pastes: Their Structure and Properties*, Cement and Concrete Association, Slough, UK, pp. 2–30.

Diamond, S. (2000) "Microscopic features of ground water-induced sulfate attack in highly permeable concretes", presented at the ACI/CANMET mtg., Barcelona, Spain, June.

Diamond, S. (2001) "Calcium hydroxide in cement paste and concrete – a microstructural appraisal", in J. Skalny, J. Gebauer and I. Odler (eds) *Materials Science of Concrete Special Volume: Calcium Hydroxide in Concrete* (in press).

Diamond, S. and Lee, R.J. (1999) "Microstructural alterations associated with sulfate attack in permeable concretes", in J. Marchand and J. Skalny (eds) *Materials Science of Concrete Special Volume: Sulfate Attack Mechanisms*, The American Ceramic Society, Westerville, OH, pp. 123–173.

Diamond, S., Olek, J. and Wang, Y. (1998) "The occurrence of two-tone structures in room-temperature cured cement paste", *Cem. Concr. Res.* **28**: 1237–1243.

Dimic, D. and Drolc (1986) "The influence of alite content on the sulfate resistance of portland cement", *Proceedings of the 8th International Congress on the Chemistry of Cement*, Rio de Janeiro, **5**: 195–199.

Djuric, M., Ranogajec, R., Omorjan, R. and Niletic, S. (1996) "Sulfate corrosion of portland cement – pure and blended with 30% of fly ash", *Cement and Concrete Research* **26**: 1295–1300.

Eglinton, M. (1998) "Resistance of Concrete to Destructive Agents", in P.C. Hawlett (ed.), *Lea's Chemistry of Cement and Concrete*, Arnold, London, pp. 300–342.

Erlin, B. (1996) "Ettringite – whatever you may think it is", in *Proc. 18th Int. Conf. Cement Microscopy*, ICMA, pp. 380–381.

Erlin, B. (ed.) (1999) *Ettringite: The Sometimes Host of Destruction*, ACI SP-177, 16 presentations, 265 pp.

Famy, C. (1999) "Expansion of heat-cured mortars", Ph.D Thesis, University of London, September.

Federal Supplement (1995) Lone Star Industries Inc. Concrete Railroad Cross Ties Litigation, Federal Supplement 822 (D.Md. 1995).

Ferrari, F. (1919) German Patent # 394 005, "Langsambindender Eisenoxydreicher Gesinterter Zement", 6 July 1920.

Ferraris, C.F., Clifton, J.R., Stutzman, P.E. and Gaboczi, E.J. (1997) "Mechanism of degradation of Portland cement-based systems by sulfate attack", in K. Scrivener and J.F. Young (eds) *Mechanism of Chemical Degradation of Cement-based Systems*, E & FN Spon, London, pp. 185–172.

Folliard, K.J. and Sandberg, P. (1994) "Mechanism of concrete deterioration by sodium sulphate crystallization", *International Conference on Durability of Concrete*, Nice France (SP-145), pp. 933–945.

Fu, Y. and Beaudoin, J.J. (1996) "Microcracking as a precursor to delayed ettringite formation in cement systems", *Cem. Concr. Res.* **26**: 1493–1498.

Fu, Y., Xie, P., Gu, P. and Beaudoin, J.J. (1993) "Preferred nucleation of secondary ettringite in preexisting cracks of steam cured cement paste", *Journal of Mater. Sci. Letters* **12**(23): 1864–1865.

Fu, Y., Xie, P., Gu, P. and Beaudoin, J.J. (1994) "Significance of preexisting cracks on nucleation of secondary ettringite in steam cured cement paste", *Cem. Concr. Res.* **24**: 1015–1024.

Gartner, E.M. and Gaidis, J.M. (1989) "Hydration Mechanisms I", in J. Skalny (ed.) *Materials Science of Concrete* I, The American Ceramic Society, Westerville, OH, pp. 95–125.

Gaze, M.E. (1997) "The effects of varying gypsum content on thaumasite formation in a cement-lime-sand-mortar at 5 °C", *Cement and Concrete Research* **27**: 259–265.

Gebhardt, R.F. (1995) "Survey of North American Portland cements: 1994", *Cement, Concrete, and Aggregate*, **17**(2): 145.

Giergiczny Z. (1997) "Sulphate resistance of cements with mineral admixtures", *Proceedings of the 10th International Congress on the Chemistry of Cement*, Gothenburg, paper 4iv019.

Glasser, F.P. (1996) "The role of sulfate mineralogy and cure temperature in delayed ettringite formation", *Cement and Concrete Composites* **8**: 187–193.

Glasser, F.P. (1999) "Reactions between cement paste and sulfate ions", in J. Marchand and J. Skalny (eds) *Materials Science of Concrete Special Volume: Sulfate Attack Mechanisms*, The American Ceramic Society, Westerville, OH, pp. 99–122.

Gollop, R.S. and Taylor, H.F.W. (1992–1996) "Microstructural and microanalytical studies of sulfate attack I. Ordinary cement paste. II. Sulfate resisting Portland cement: Ferrite composition and hydration chemistry. III. Sulfate-resisting Portland cement: Reactions with sodium and magnesium sulfate solutions. IV. Reactions of a slag cement paste with sodium and magnesium sulfate solutions. V. Comparison of different slag blends", *Cement and Concrete Research* **22**: 1027–1038; **24**: 1347–1358; **25**: 1581–1590; **26**: 1013–1028; **26**: 1029–1044.

Gospodinov, P.N., Kazandjiev, R.F., Partalin, T.A. and Mironove, M.K. (1999) "Diffusion of sulfate ions into cement stone regarding simultaneous chemical reaction and resulting effects", *Cement and Concrete Research* **29**: 1591–1596.

Grattan-Bellew P.E., Beaudoin, J.J. and Vallee, V.-G. (1998) "Effect of agggregate particle size and composition on expansion of mortar bars due to delayed ettringite formation", *Cem. Concr. Res.* **28**: 1147–1156.

Guerrero, A., Goni, S. and Macias, A. (2000) "Durability of new fly ash-belite cement mortars in sulfated and chloride medium", *Cement and Concrete Research* **30**: 1231–1238.

Halliwell, M.A. and Crammond, N.J. (1996) "Deterioration of brickwork retaining walls as a result of thaumasite formation" in C. Sjorstrom (ed.) *Durability of Building Materials and Components 7*, vol. 1, E & FN Spon, London, pp. 235–244.

Halliwell, M.A. and Crammond, N.J. (1999) "Avoiding the thaumasite form of sulfate attack", *Proceedings 8th DBMC*, Vancouver, Canada.

Halliwell, M.A., Crammond, N.J. and Barker, A.P. (1996) "The thaumasite form of sulfate attack in limestone-filled cement mortars", BRE Lab. Report BR 307, Construction Research Communications Ltd. (CRC), UK.

Hansen, W.C. (1963) "Crystal growth as a source of expansion in Portland cement concrete", *Proceedings of the American Society for Testing and Materials* **63**: 932–945.

Haynes, H. (2000) "Sulfate attack on concrete: Laboratory versus field experience", *Suppl. Proc. 5th CANMET/ACI Int. Conf. Durability of Concrete*, Barcelona, June (in press).

Haynes, H., O'Neill, R. and Mehta, P.K. (1996) "Concrete deterioration from physical attack by salts", *Concrete International* **18** (1 January): 63–68.

Hearn, N. and Figg, J. (2000) "Transport mechanisms and damage: current issues in permeation characteristics of concrete", in S. Mindess and J. Skalny (eds) *Materials Science of Concrete*, vol. VI, The American Ceramic Society, Westerville, OH (in press).

Heinz, D. (1986) Doctoral Thesis, Technical University of Rhein-Westfalen, Aachen, Germany.

Heinz, D. and Ludwig, U. (1986) "Mechanisms of subsequent ettringite formation in mortars and concretes after heat treatment", *Proc. 8th Int. Conf. Chem. Cem.*, vol. 5, Rio de Janeiro, p. 189.

Heinz, D. and Ludwig, U. (1987) "Mechanism of secondary ettringite formation in mortars and concretes subjected to heat treatment", in J.M. Scanlon (ed.) *Concrete Durability: Katherine and Bryant Mather International Conference* ACI SP-100, **2**: 2059–2071.

Heinz, D., Ludwig, U. and Rudiger, I. (1989) "Delayed ettringite formation in heat treated mortars and concrete", *Concrete Precasting Plant and Technology*, pp. 56–61.

Herr, R., Wieker, W. and Winkler, A. (1988) "On the chemical equilibria of hydroxide and sulfate ion concentrations in the pore solution of thermal treated concrete", *in Proc. 5th Natl. Conf. on Mechanics and Technology of Composite Materials*, Varna, Bulgaria, 29 September–1 October, pp. 550–555.

Herrick, J., Scrivener, K.L. and Pratt, P.L. (1992) "The development of microstructure in calcium sulfoaluminate expansive cement", *Materials Research Society Symposium Proceedings* **245**: 277–282.

Hime, W.G. (1996a) "Clinker sulfate: a cause for distress and a need for specification", in R.K. Dhir and T.D. Dyer (eds) *Concrete for Environment Enhancement and Protection*, London: E&FN Spon, pp. 387–395.

Hime, W.G. (1996b) "Delayed ettringite formation", in *Proc. 18th Int. Conf. Cement Microscopy*, ICMA, p. 378.

Hime, W.G. and Mather, B. (1999) "'Sulfate attack' or is it?", *Cement and Concrete Research* **29**: 789–791.

Hobbs, D.W. (1999) "Expansion and cracking in concrete associated with delayed ettringite formation", in B. Erlin (ed.) *Ettringite – The Sometimes Host of Destruction*, ACI SP **177**: 159–181.

Hobbs, D.W. and Taylor, H.F.W. (2000) "Nature of the thaumasite sulfate attack mechanism in field concrete", *Cement and Concrete Research* **30**: 529–534.

Hormann, K., Hofmann, F.J. and Schmidt, M. (1997) "Stability of concrete against biogenic sulfuric acid corrosion – a new method for determination", *Proceedings of the 10th International Congress on the Chemistry of Cement*, Gothenburg, paper 4iv038.

Hulshizer, A.J. (1997) "Air-entrainment control or consequences" *Concrete International* (July): 38–39.

Irassar, D.F. and Ans-Batic, O. (1989) "Effect of low calcium fly ash on sulfate resistance of Portland cement", *Cement and Concrete Research* **19**: 194–202.

Jie, Y., Warner, N., Clark, B.A., Thaulow, D.A. and Skalny, J. (1993) "Temperature relics in steam cured concrete", in *Proc. Inc. Cem. Microscopy Assoc.*, pp. 289–299.

Johansen, V., Thaulow, N., Jakobsen, U.H. and Palbol, L. (1993) "Heat-curing induced expansion", *Proc. 3rd Beijing Int. Symp. on Cement and Concrete*, p. 144.

Johansen, V., Thaulow, N. and Skalny, J. (1993) "Simultaneous presence of alkali–silica gel and ettringite in concrete", *Adv. Cem. Res.* **5**: 23–29.

Johansen, V., Thaulow, N. and Skalny, J. (1995) "Internal reactions causing cracking of concrete", *Beton + Fertigteil Technik* **11**: 56–68.

Johansen, V. and Thaulow, N. (1999) "Heat curing and late formation of ettringite", in B. Erlin (ed.) *Ettringite: The Sometimes Host of Destruction*, ACI SP-177, pp. 47–64.

Kalousek, G.W. and Benton, E.J. (1970) "Mechanism of sea water attack on cement pastes", *Journal of the American Concrete Institute* **67**: 187–192.

Kantro, D.L., Brunauer, S. and Weise, C.H. (1963) "Development of surface in the hydration of calcium silicates", *J. Phys. Chem.* **66**: 1804.

Kelham, S. (1996) "The effect of cement composition and fineness on expansion associated with delayed ettringite formation", *Cement and Concrete Composites* **18**: 171–179.

Kelham, S. (1999) "The influence of cement composition on the volume stability of mortar", in B. Erlin (ed.) *Ettringite – The Sometimes Host of Destruction*, ACI SP-177, 27–45.

Kjellsen, K.O., Detwiller, R.J. and Gjorv, O.E. (1991) "Development of microstructures in plain cement pastes hydrated at different temperatures", *Cem. Concr. Res.* **21**: 179–189.

Klemm, W.A and Miller, F.M. (1997) "Plausibility of delayed ettringite formation as a distress mechanism – considerations at ambient and elevated temperature", in H. Justnes (ed.) *Proc. 10th Int. Congress on the Chemistry of Cement*, vol. 4, Gothenburg, Sweden, paper 4iv059.

Krizan, D. and Zivanovic, B. (1997) "Resistance of fly ash cement mortars to sulphate attack", *Proceeding of the 10th International Congress on Cement Chemistry*, Gothenburg, paper 4iv020.

Lafuma, H. (1927) "Theory of expansion of cement" (in French), *Revue des Materiaux de Construction et de Traveaux Public* **243**: 441–444.

Lawrence, C.D., Myers, J.J. and Carrasquillo, R.L. (1999) "Premature concrete deterioration in Texas department of transportation precast elements", in B. Erlin (ed.) *Ettringite – The Sometimes Host of Destruction*, ACI, SP-177, pp. 141–158.

Lawrence, C.D. (1993) "Laboratory studies of concrete expansion arising from delayed ettringite formation", *Technical Report C/16*, British Cement Association.

Lawrence, C.D. (1995a) "Mortar expansions due to delayed ettringite formation", *Cem. Concr. Res.* **25**: 903–914.

Lawrence C.D. (1995b) "Delayed ettringite formation: an issue?", in J. Skalny and S. Mindess (eds) *Material Science of Concrete*, vol. IV, The American Ceramic Society, Westerville, OH, p. 113.

Lawrence, C.D. (1999) "Long-term expansion of mortars and concretes", in B. Erlin, (ed.) *Ettringite – The Sometimes Host of Destruction*, ACI SP-177, 105–123.

Lewis, M.C. (1996) Heat curing and delayed ettringite formation in Concrete, Ph.D Thesis, Imperial College, Materials Department, London.

Lewis, M.C., Scrivener, K.L. and Kelham, S. (1995) "Heat curing and delayed ettringite formation", *MRS Symposia Proceedings* **370**: 67–76.

Li, G., Le Bescop, P. and Moranville, M. (1996) "The U-Phase formation in cement-based systems containing high amounts of $Na_2SO_4$", *Cement and Concrete Research* **26**: 27–33.

Li, G., Le Bescop, P. and Moranville, M. (1996) "Expansion mechanism associated with the secondary formation of the U-phase in cement-based systems containing high amounts of $Na_2SO_4$", *Cement and Concrete Research* **26**: 195–201.

Lota, L.S, Pratt, P.L. and Bensted, J. (1995) "A discussion of the paper 'Microstructural and microanalytical studies of sulfate attack' by R.S. Gollop and H.F.W. Taylor", *Cement and Concrete Research* **25**: 1811–1813.

Ludwig, U. (1991) "Problems of ettringite reformation in heat treated mortars and concretes", in *11th IBAUSIL Proceedings*, Weimar, pp. 164–177.

Mangat, P.S. and Khatib, J.M. (1992) "Influence of initial curing on sulphate resistance of blended cement concrete", *Cement and Concrete Research* **22**: 1089–1100.

Mangat, P.S. and Khatib, J.M. (1995) "Influence of fly ash, silica fume and slag on sulfate resistance of concrete", *ACI Materials Journal* **92**: 542–552.

Marchand, J. and Skalny, J. (eds) (1999) *Materials Science of Concrete Special Volume: Sulfate Attack Mechanisms*, The American Ceramic Society, Westerville, OH, 371pp.

Marusin, S.L. (1993) "SEM studies of DEF in hardened concrete", in *Proc. 15th Int. Conf. Cement Microscopy*, Dallas, TX, pp. 289–299.

Marusin, S.L. and Hime, W.G. (1996) "Concrete and delayed ettringite formation – causes, development and case studies", *Supplementary papers of the 4th CANMET/ACI Int. Conference On Durability of Concrete*, Sydney, Australia, pp. 1–12.

Marks, V.J. and Dubberke, W.G. (1996) "A different perspective for investigation of Portland cement concrete deterioration", *Transportation Research Record No. 1525*.

Mather B. (1973) "A discussion of the paper 'Mechanism of expansion associated with ettringite formation' by P.K. Mehta", *Cement and Concrete Research* **3**: 651–652.

Mather, B. (1993) Memorandum for Record, Waterways Experiment Station, Vicksburg, MI, 16 April.

Mather, B. (1996) "Discussion of 'The process of sulfate attack on cement mortars'", *Advances in Cement Based Materials* **5**: 109–110.

Mehta, P.K. (1973a) "Mechanism of expansion associated with ettringite formation", *Cement and Concrete Research* **3**: 1–6.

Mehta, P.K. (1973b) "Effect of lime on hydration of pastes containing gypsum and calcium aluminates or calcium sulfoaluminates", *Journal of the American Ceramic Society* **56**: 315–319.

Mehta, P.K. (1991) *Concrete in Marine Environment*, Elsevier Applied Science, London.

Mehta, P.K. (1992) "Sulfate attack on concrete – a critical review" in J. Skalny (ed.) *Material Science of Concrete*, American Ceramic Society, Westerville, OH, pp. 105–130.

Mehta, P.K. (1999) "Sulfate attack in marine environment", in J. Marchand and J. Skalny (eds) *Materials Science of Concrete Special Volume: Sulfate Attack Mechanisms*. The American Ceramic Society, Westerville, OH, 1999, pp. 295–300.

Mehta, P.K. (2000) "Sulfate attack on concrete: Separating the myth from reality", *Concrete International* **22**(8): 57–61.

Mehta, P.K. and Hu, F. (1978) "Further evidence for expansion of ettringite by water absorption", *Journal of the American Ceramic Society* **61**: 179–181.

Mehta, P.K. and Haynes, H. (1975) "Durability of concrete in seawater", *Journal of the American Society of Civil Engineers Structural Division* **101**(ST8): 1679–1686.

Mehta, P.K. and Monteiro, P. (1993) *Concrete*, 2nd edn, McGraw-Hill.

Mehta, P.K., Pirtz, D. and Polivka, M. (1979) "Properties of alite cement", *Cement and Concrete Research* **9**: 439–450.

Mehta, P.K. and Wang, S. (1982) "Expansion of ettringite by water absorption", *Cement and Concrete Research* **12**: 121–122.

Michaelis, W. (1901) German Patent # 143 604, "Verfahren zur herstellung eines dem Meerenswasser wiederstehenden Zements aus Kalk und Eisenoxyd o. dgl.", 12 February.

Michaud, V. and Suderman, R. (1997) "The solubility of sulfates in high $SO_3$ clinkers," in B. Erlin (ed.) "Ettringite – the sometime host of destruction", ACI SP-177, pp. 15–25.

Mielenz, R.O., Marusin, S.L., Hime, W.G. and Jugovic, Z.T. (1995) "Investigation of prestressed concrete railway tie distress", *Concrete International* **17**: 62–68.

Miletic, S.R. and Ilic, M.R. (1997) "Effect of ammonium sulphate corrosion on the strength of concrete", *Proceedings of the 10th International Congress on the Chemistry of Cement*, Gothenburg, paper 4iv023.

Miller, G. and Tang, F. (1996) "The distribution of sulfur in present-day clinkers of variable sulfur content", *Cement and Concrete Research* **26**(12): 1821–1829.

Moldovan, V. and Butucescu, N. (1980) "Expansion mechanism of expansive cement", *Proceedings of the 7th International Congress on the Chemistry of Cement*, Paris, **3**: V-1–V-5.

Montani, S. (1997) "Delayed ettringite formation – literature review and tests at HMC", *Holderbank Management Center – Report*, PDA 97/17'014/E, October, 23 pp.

Monteni, J., Vincke, E., Beeldens, A., De Belie, N., Taerwe, L., Van Gemert, D. and Verstraete, W. (2000) "Chemical, microbiological and *in situ* test methods for biogenic sulfuric acid corrosion of concrete", *Cement and Concrete Research* **30**: 623–634.

Moranville, M. and Li, G. (1999) "The U-phase – formation and stability", in J. Marchand and J. Skalny (eds) *Materials Science of Concrete; Sulfate Attack Mechanism*, The American Ceramic Society, Westerville OH, pp. 175–188.

Oberholster, R.E., van Aardt, J.H.P. and Brandt, M.P. (1983) "Durability of cementitious systams" in P. Barnes (ed.) *Structure and Performance of Cements*, London and New York: Applied Science publishers, pp. 365–413.

Oberholster, R.E., Maree, H. and Brand, J.H.B. (1992) "Cracked prestressed concrete railway sleepers: Alkali–silica reaction or delayed ettringite formation," in *Proceedings of the 9th International Conference of Alkali-Aggregate Reaction in Concrete*, ACI, **2**: 739–749.

Odler, I. (1997) "Ettringite nomenclature (letter to the Editor)", *Cem. Concr. Res.* **27**(3): 473–474.

Odler, I. (2001) "Free lime content and unsoundness of cement", in J. Skalny, J. Gebauer and I. Odler (eds) *Materials Science of Concrete Special Volume: Calcium Hydroxide in Concrete* (in press).

Odler, I. and Chen, Y. (1995) "Effect of cement composition on the expansion of heat-cured cement paste," *Cem. Concr. Res.* **25**: 853–862.

Odler, I. and Chen, Y. (1996) "On the delayed expansion of heat cured Portland cement pastes and concretes", *Cement and Concrete Composites* **18**: 181–185.

Odler, I. and Collan-Subauste, J. (1999) "Investigation on cement expansion associated with ettringite formation", *Cement and Concrete Research* **29**: 731–735.

Odler, I. and Gasser, M. (1988) "Mechanism of sulfate expansion in hydrated Portland cement", *Journal of the American Ceramic Society* **71**: 1015–1020.

Odler, I. and Jawed, I. (1991) "Expansive reactions in concrete" in J. Skalny and S. Mindess (eds) *Materials Science of Concrete*, The American Ceramic Society, Westerville, OH, pp. 221–247.

Odler, I. and Skalny, J. (1973) "Hydration of $C_3S$ at elevated temperatures", *Appl. Chem. Biotechnol.* **23**: 661–667.

Ogawa, K. and Roy, D.M. (1981–1982) "Hydration, ettringite formation and its expansion mechanism", *Cement and Concrete Research* **11**: 741–750; **12**: 101–109.

Ouyang, C. and Lane, O.J. (1997) "Freeze-thaw durability of concretes with infilling of ettringite in voids", Iowa DOT Report for research project MLR 94-4, Phase 2.

Pade, C., Jakobsen, U.H. and Johansen, V. (1997) "Evolution of expansion, weight gain, microstructure, and microchemistry of DEF-affected mortar", *Proc. 13th Int. Baustofftagung*, Bauhaus-Universitat, Weimar, Band 1, 1-0521–1-0533.

Parker, C.D. (1945) "The corrosion of concrete: 1. The isolation of a species of bacterium associated with the corrosion of concrete exposed to atmospheres containing hydrogen sulphide", *The Australian Journal of Experimental Biology and Medical Science* **23**: 81–90.

Pettifer, K. and Nixon, P.J. (1980) "Alkali metal sulphate – a factor common to both alkali aggregate reaction and sulphate attack on concrete", *Cement and Concrete Research* **10**: 173–181.

Ping, X. and Beaudoin, J.J. (1992) "Mechanism of sulfate expansion, I. Thermodynamic principles of crystallization pressure, II. Validation of thermodynamic theory", *Cement and Concrete Research* **22**: 631–640, 845–854.

PCA (1996) "Portland cement: past and present characteristics", *Concrete Technology Today*, July.

Quyang, C. and Lane, O.J. (1997) "Effect of infilling air voids by ettringite on freeze–thaw durability of concrete", Presentation at the ACI Spring Convention, Seattle, April.

Regourd, M., Hornain, H. and Mortureux, B. (1980) "Microstructure of concrete in aggressive environments", in *Durability of Building Materials and Components*, ASTM, Philadelphia, PA, pp. 253–268.

RILEM (2001) *Internal Sulfate Attack – RILEM Report of TC-ISA*, J. Skalny and K. Scrivener (eds) (in preparation).

Roberts, L.L. and Hooton, D. (1997) (Organizers) Symposium on Internal Sulfate Attack on Cementitious Systems: Implications and Standards, ASTM Committee C-1, San Diego, December.

Rodrigez-Navarro, C., Doehne, E. and Sebastian. H. (2000) "How does sodium sulfate crystallize? – Implications for the decay and testing of building materials", *Cement and Concrete Research* **30**: 1527–1534.

Scherer, G.W. (1999) "Crystallization in pores", *Cement and Concrete Research* **29**: 1347–1358.

Scrivener, K.L. (1989) "The microstructure of concrete", in J. Skalny (ed.) *Materials Science of Concrete I*, vol. I, The American Ceramic Society, pp. 127–161.

Scrivener, K.L. (1992) "The effect of heat treatment on inner product C-S-H", *Cem. and Concr. Res.* **22**: 1224–1226.

Scrivener, K.L. (1996) "Delayed ettringite formation and concrete railroad tiles", in *Proc. 18th Int. Conf. Cement Microscopy*, ICMA, pp. 375–377.

Scrivener, K.L. (2000) personal communication to J. Skalny.

Scrivener, K.L. and Capmas, A. (1998) "Calcium aluminate cements", in P.C. Hawlett (ed.) *Lea's Chemistry of Cement and Concrete*, Arnold, London, pp. 709–778.

Scrivener, K.L. and Taylor, H.F.W. (1993) "Delayed ettringite formation: a microstructural and microanalytical study", *Advances in Cement Research* **5**: 139–146.

Sersale, R., Cioffi, R., de Vito, B., Frigione, G. and Zenone, F. (1997) "Sulphate attack on carbonated and uncarbonated Portland and blended cements", *Proceedings of the 10th International Congress of the Chemistry of Cement*, Gothenburg, paper 4iv17.

Sibbick, R.G. and Crammond, N.N. (1999) "Two case studies into the development of the thaumasite form of sulfate attack (TSA) in hardened concretes", *Proceedings 7th Euroseminar on Microscopy Applied to Building Materials, Delft, Holland*.

Shayan, A. and Quick, G.W. (1991–1992) "Relative importance of deleterious reactions in concrete: formation of alkali-aggregate reactivity products and secondary ettringite", *Advances in Cement Research* **4**(16): 149–157.

Siedel, H., Hempel, S. and Hempel, R. (1993) "Secondary ettringite formation in heat treated portland cement concrete: influence of different w/c ratios and heat treatment temperatures", *Cem. Concr. Res.* **23**: 453–461.

Skalny, J. (1994) "Evaluation of concrete cores", Report to Portland Cement Association.

Skalny, J., Diamond, S. and Lee, R.J. (1998) "Sulfate attack, interfaces and concrete deterioration", in A. Katz, A. Bentur, M. Alexander and G. Arliguie (eds) *Proceedings, RILEM 2nd International Conference on The Interfacial Transition Zone in Cementitious Composites*, NBRI Technion, Haifa, pp. 141–151.

Skalny, J., Johansen, V. and Miller, F.M. (1997) "Sulfates in cement clinker and their effect on concrete durability", in V.M. Malhotra, (ed.) *Proc. 3rd CANMET/ACI Int. Symp. On Advances in Concrete Technology*, ACI SP 171–30, **2**: 625–631.

Skalny, J., Johansen, V., Thaulow, N. and Palomo, A. (1996) "DEF as a form of sulfate attack", *Materiales de Construccion* **46**: 244, 5–29.

Skalny, J. and Klemm, W.A. (1981) "Alkalis in clinker: origin, chemistry, effects", *Proc. Conf. Alkali-aggregate Reaction in Concrete*, Cape Town, CSIR, S252/1-7.

Skalny, J. and Locher, F.W. (1999) "Curing practices and delayed ettringite formation: the European experience", *Cem. Concr. Aggr.* **21**: ASTM, June.

Skalny, J. and Odler, I. (1972) "Pore structure of hydrated calcium silicates: III. influence of temperature on the pore structure of hydrated tricalcium silicate", *J. Colloid Interface Sci.* **40**: 199.

Skalny, J., Odler, I. and Young, F. (2000) "Discussion of the paper 'Sulfate attack,' or is it" by W.G. Hime and B. Mather, *Cement and Concrete Research* **30**: 161–162.

Skalny, J. and Pierce, J. (1999) "Sulfate attack: an overview", in J. Marchand and J. Skalny (eds) *Materials Science of Concrete Special Volume: Sulfate Attack Mechanisms*, The American Ceramic Society, Westerville, OH, pp. 49–63.

Soroushian P. and Alhozami A. (1992) "Correlation between fly ash effects on permeability and sulfate resistance of concrete", *Proceedings of the 9th International Congress on the Chemistry of Cement*, New Delhi, **5**: 196–202.

Stark, J. and Bollmann (1997) "Ettringite formation – a durability problem of concrete pavements", in *Proc. 10th Intern. Conrg. Chem. Cem.*, Gothenburg, Sweden, vol. IV, paper4iv062, 8 pp.

Stark, J. and Bollmann (2000) "Late ettringite formation in concrete, Part 2", *ZKG International* **53**(4): 232–240.

Swenson, E.G. (ed.) (1968) *Performance of Concrete*, University of Toronto Press, Toronto.

Sylla, M.-H. (1988) "Reactions in cement stone due to heat treatment", *Beton* **38**: 449–454.

Taylor, H.F.W. (1994a) "Delayed ettringite formation", in M.W. Grutzek and S.L. Sarkar (eds) *Advances in Cement and Concrete*, American Society of Civil Engineers.

Taylor, H.F.W. (1994b) "Sulfate reactions in concrete – microstructural and chemical aspects", in E. Gartner and H. Uchikawa (eds) *Cement Technology*, The American Ceramic Society, Westerville, OH, pp. 61–78.

Taylor, H.F.W. (1996) "Ettringite in cement paste and concrete", presentation at the *Beton:du Materiau a la Structure* conference, Arles, France.

Taylor, H.W.F. (1997) *Cement Chemistry*, 2nd edn, Thomas Telford, London.

Taylor, H.F.W. (2000a) Presentation at the Society of Chemical Industry, London, 24th February.

Taylor, H.F.W. (2000b) "Delayed ettringite formation", *The 2000 Della M. Roy Lecture*, Annual Meeting of the American Ceramic Society, Cinncinnatti, OH, May.

Taylor, H.F.W. (2000c) "Ettringite: friend or foe?," presentation at the Workshop on the Role of Calcium Hydroxide in Concrete, Holmes Beach, FL, 3 November 2000; to be published in Skalny *et al.* (eds) *Materials Science of Concrete Special Volume: Calcium Hydroxide in Concrete*, The American Ceramic Society, 2001.

Taylor, H.F.W., Famy, C. and Scrivener, K.L. (2001) "Delayed ettringite formation", *Cement and Concrete Research* **31** (in press).

TEG Report (1999) "The thaumasite form of sulfate attack: risks, diagnosis, remedial works and guidance on new construction", Report of the Thaumasite Expert Group, Department of the Environment, Transport and the Regions.

Tennis, P.D., Bhattacharja, S., Klemm, W.A. and Miller, F.M. (1997) "Assessing the distribution of sulfate in portland cement and clinker and its influence on expansion in mortar", presented at the *ASTM Symposium on Internal Sulfate Attack on Cementitious Systems: Implications for Standards Development*, San Diego, December.

Tian, B. and Cohen M.D. (2000) "Does gypsum formation during sulfate attack on concrete lead to expansion?", *Cement and Concrete Research* **30**: 117–123.

Thaulow, N. (1987) "Microscopy – the inside story of concrete", 1st Euroseminar on Microscopy Applied to Building Materials, Copenhagen, June.

Thomas, M.D.A. (1998) "Delayed ettringite formation in concrete: Recent developments and future directions", University of Toronto, 1998; to be published in S. Mindess and J. Skalny (eds) *Materials Science of Concrete*, vol. VI, The American Ceramic Society, Westerville, OH (in press).

Thomas, M.D.A. (2000) "Ambient temperature DEF: Weighing the evidence", presented at the ACI/CANMET mtg. in Barcelona, Spain, June.

Thorvaldson, T. (1952) "Chemical aspects of the durability of cement products", *Proc. 3rd International Symposium on the Chemistry of Cements*, London, UK, pp. 436–484.

Thorvaldson, T., Harris, R.H. and Wolochov, D. (1925) "Disintegration of Portland cement in sulfate waters", *Industrial and Engineering Chemistry* **17**(3): 467–470.

Thorvaldson, T., Lamour, R.K. and Vigfusson, V.A. (1928) "The expansion of Portland cement mortar bars during disintegration in sulphate solutions", *The Engineering Journal* **10**(4): 199–206.

Thorvaldson, T., Vigfusson, V.A. and Larmour, R.K. (1927) "The action of sulfates on the components of portland cement", *Trans. Royal. Soc. Canada*, 3rd Series, **21**, Section III, 295.

Wang J.G. (1994) "Sulfate attack on hardened cement paste", *Cement and Concrete Research* **24**: 735–742.

Wang, S., Ji, S., Wang, H. and Zhou, M. (1985) "Experiments on the mechanism of ettringite expansion", *Proceedings of the 1985 Beijing International Symposium on Cement and Concrete*, vol. 3.

Werner, K.-C., Chen, Y. and Odler, I. (2000) "Investigation on stress corrosion of hydrated cement pastes", *Cement and Concrete Research* **30**: 1443–1452.

Wieker, W. and Herr, R. (1989) "Zu einigen Problemen der Chemie des Portlandzements", *Z. Chemie* **29**: 321–327.

Wig, R.J. and Williams, G.M. (1915) "Investigations on the durability of cement drain tiles in alkali soils", *Technological Papers of the Bureau of Standards*, No. 44, Washington, DC: US Government Printing Office, 56 pp.

Wig, R.J., Williams, G.M. and Finn, A.N. (1917) "Durability of cement drain tiles and concrete in alkali soils", *Technological Papers of the Bureau of Standards*, No. 95, Washington, DC: US Government Printing Office, 95 pp.

Yang, R., Lawrence, C.D., Lyndsdale, C.J. and Sharp, J.H. (1999a) "Delayed ettringite formation in heat-cured portland cement mortars", *Cem. Concr. Res.* **29**: 17–25.

Yang, R., Lawrence, C.D. and Sharp, J.H. (1999b) "Effect of type of aggregate on delayed ettringite formation", *Adv. Cem. Res.* **11**(2): 1–14.

Yang, Q., Wu, X. and Huang, S. (1997) "Concrete deterioration due to physical attack by salt crystallization, *Proc. 10th Int. Congress on the Chemistry of Cement*, Gothenburg, Sweden, SINTEF, paper 4iv032.

Yang, S., Zhonghi, X. and Mingshu, T. (1996) "The process of sulfate attack on cement mortars", *Advances in Cement Based Materials* **4**: 1–5.

Young, J.F. (1998) "Sulfate attack" (letter to the Editor), *Concrete International* **20**(8): 7.

Zhaozhou Zhang (1999) *Delayed Ettringite Formation in Heat Cured Cementitious Systems*, Ph.D Thesis, Purdue University, December.

# 5 Consequences of sulfate attack on concrete

## 5.1 INTRODUCTION

As discussed in the previous chapter, concrete subject to sulfate attack undergoes a progressive and profound reorganization of its internal microstructure. These alterations have direct consequences on the engineering properties of the material. As will be seen in the following paragraphs, concrete undergoing sulfate attack is often found to suffer from swelling, spalling and cracking. There is overwhelming evidence to show that the degradation also contribute to significantly reduce the mechanical properties of concrete. Many structures affected by sulfate degradation often need to be repaired or, in the most severe cases, partially reconstructed.

The various consequences of sulfate attack on concrete are reviewed in the following sections. Throughout the text, distinction is made between concrete suffering from internal sulfate degradation and that affected by external sulfate attack. The behavior of hydrated cement systems tested under well-controlled laboratory conditions is also distinguished from the performance of concrete in service.

## 5.2 EXTERNAL APPEARANCE AND VOLUME STABILITY OF CONCRETE ATTACKED BY SULFATE

As emphasized in Chapter 4, it is now well established that hydrated cement systems subject to sulfate attack often sustain damage as a result of excessive volume change. For instance, the swelling behavior of concrete suffering from "internal" sulfate attack has been the subject of numerous reports (Day 1992; Diamond 1996; Lawrence 1995a,b). The volume instability of mortar and concrete mixtures exposed sulfate solutions is also well documented (Gollop and Taylor 1992–1996, Mehta 1992; Odler and Jawed 1991; Thorvaldson 1952). For additional information see Thorvaldson et al. (1925, 1927, 1928) and Tuthill (1936).

The macroscopic manifestations of both types of degradation will be reviewed in separate sections. The volume instability of concrete subjected to internal sulfate attack will be briefly reviewed in Section 5.2.1. The topic has already been discussed in the previous chapter (see Section 4.8.2). A more comprehensive discussion of the macroscopic behavior of concrete suffering from external sulfate attack is presented in Section 5.2.2.

### 5.2.1   External appearance and volume stability of concrete subjected to internal sulfate attack

Deleterious expansion may occur in concrete when excessive amounts of gypsum or anhydrite, in quantities well above normal levels, are present in the cement. Excess sulfate in concrete can also originate from contaminated aggregates (Figg 1999; St John *et al.* 1958). As emphasized by Harboe (1982), admissible limits for the aggregate sulfate content may vary from one source to another, and laboratory trial tests are recommended to evaluate the performance of potentially reactive aggregates.

Typical dilation curves for laboratory samples subjected to internal sulfate attack are shown in Figure 5.1 (Ouyang *et al.* 1988). These results were obtained by testing a series of mortar prisms according to the prescriptions of ASTM C452. It should be emphasized that, in this case, expansion arised from excess gypsum initially added to the various mixtures during their production. In that respect, the curves appearing in Figure 5.1 are examples of what has been referred to as *composition-induced internal sulfate attack* in Section 4.8.2.

As can be seen in Figure 5.1, expansion develops very quickly. Furthermore, the rate of expansion becomes linear after only a few days of test. Ouyang *et al.* (1988) reported that significant cracking and strength loss could be observed when the samples had experienced approximately 0.3% expansion. This aspect of the problem will be further discussed in Section 5.4.

ASTM C1038 allows the quantity of $SO_3$ in the cement and pozzolan, and thus in concrete, to increase beyond the chemical requirements by any amounts, as long as expansion after fourteen days of immersion in water is less than 0.02%. Recently, Day (2000) found a positive correlation between the fourteen-day dilation measured according to ASTM C1038 and the long-term expansion of mortar.

In the literature, various other expansion limits have been proposed for concrete samples subjected to composition-induced internal sulfate attack. For instance, Samarai (1976) recommended 0.1% expansion as a safe margin for determining the maximum percentage of sulfate that can be added to a given mixture without causing any significant degradation. Crammond (1984) used 0.1% expansion after six months as the limit above which swelling becomes significantly deleterious. As pointed out by Ouyang *et al.* (1988), this limit is also suggested by ASTM C227 as an acceptance criterion for alkali-aggregate reactivity.

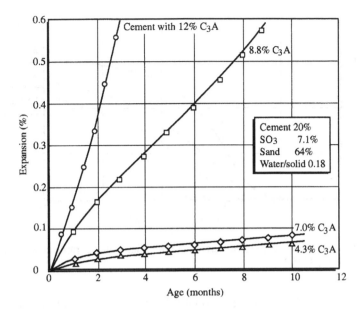

*Figure 5.1* Expansion of mortars containing different types of cement under internal sulfate attack.
*Source*: Ouyang *et al.* (1988)

According to St John *et al.* (1958), field observations of composition-induced internal sulfate attack indicate that degradation can be, in many cases, spectacular. Expansion usually occurs within weeks or months due the relatively high reactivity of gypsum. Efflorescence, scaling, spalling and cracking are widespread, and require the removal of the contaminated concrete.

A typical dilation curve for a mortar sample subjected to *heat-induced internal sulfate attack* (or DEF) is presented in Figure 5.2. As can be seen, the kinetics of expansion can be much different from what is usually seen for composition-induced internal sulfate attack (see Figure 5.1). The expansion of heat-cured samples tends to begin after an initial "incubation" period during which very little swelling if any, is observed. In addition, expansion curves are usually non-linear. As can be seen in Figure 5.2, the dilation curves are rather characterized by their S-shape.

It should also be kept in mind that a wide range of parameters might influence the kinetics of dilation of samples affected by heat-induced sulfate attack. For instance, a critical review by Day (1992) of a series of studies specifically devoted to DEF indicates that the intensity and the onset of swelling tend to vary according to the size and the shape of the samples. Laboratory experiments also demonstrate the volume instability is also affected by post-heat treatment storage conditions such as temperature, moisture content and the concentration of alkali ions in the surrounding solution (Day 1992; Famy 1999).

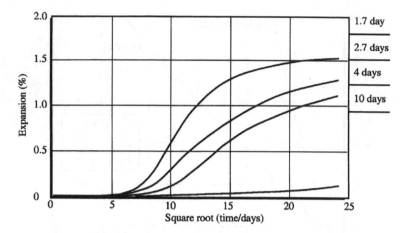

*Figure 5.2* Mortar prism expansions after extended heat cure periods at 95–100 °C.
*Source*: Lawrence (1995b)

As extensively discussed in Section 4.8.2, available information shows that heat-induced sulfate attack (or DEF) of field concrete is characterized by map cracking, longitudinal cracking and occasional warping of the element. Over the past decades, the behavior of steam-cured railroads ties suffering from DEF has received a lot of attention. In most of the documented cases, visible damage was reported several years after the products have been manufactured and in use. Damage was characterized by development of cracks that started at the corners and edges of the concrete element and gradually spread into deeper regions as the time progressed. Elements not directly exposed to moisture were usually found to be less damaged or undamaged. Typical field cases are presented in Chapter 8.

### 5.2.2　External appearance and volume stability of concrete subjected to external sulfate attack

Most of the information available on the volume instability of concrete exposed to sulfate solutions originates from laboratory experiments performed under well-controlled conditions. In a typical experiment (such as ASTM C1012), relatively small samples are kept continuously immersed in the test solution. Specimens are visually inspected at regular intervals. Changes in mass and length are also regularly monitored. A critical appraisal on the various standard test methods specifically designed to assess the durability of hydrated cement systems to sulfates is given in Chapter 9.

Typical length-change curves, obtained by Brown (1981), are given in Figure 5.3. As can be seen, the immersion of the samples in the test solutions is first followed by an "incubation" period (similar to the one seen for samples

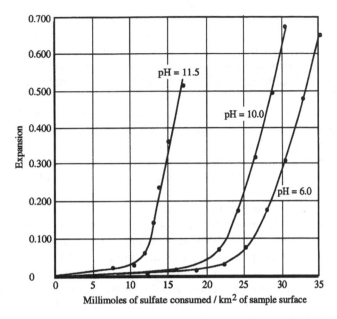

*Figure 5.3* Relationship between the mortar bar expansion and the sulfate ion consumption per unit surface area of sample at pH 6, 10 and 11.5.
*Source*: Brown (1981)

affected by heat-induced internal sulfate degradation) during which the specimens do not experience any significant swelling. However, after a few days of immersion, the length-change curve is definitively inflected upwards and finally the rate of expansion becomes almost constant until total disintegration of the samples.

Over the years, numerous expansion limits have been suggested as failure criterion for mortar and concrete samples exposed to sulfate attack. For instance, Smith (1958) defined failure as 0.5% expansion, which was found to correspond to approximately 40% loss in the dynamic modulus of elasticity. In his comprehensive review of the resistance of concrete to external sulfate attack, Tuthill (1978) used 0.4% expansion as an indication of the complete failure of test samples. More recently, Mather (1982) proposed an expansion limit of 0.1% as a criterion for failure of hydrated cement systems in contact with sulfate solutions. Patzias (1987) also suggested that 0.1% expansion at 180 days would be appropriate as a maximum acceptance limit for moderate sulfate resistance, and 0.05% expansion at 180 days as a limit for high sulfate resistance when the expansion test is performed according to ASTM C1012. The relevance of these various criteria with respect to the evolution of the mechanical properties of concrete will be further discussed in Section 5.4.

The expansion of concrete is usually accompanied by the development of cracks into the material. Numerous studies indicate that cracking is usually initiated near the surface and gradually evolves towards the central portion of the sample (Gollop and Taylor 1992–1996; Lagerblad 1999). According to Lagerblad (1999), cracks can be detected by a visual inspection when the linear expansion of the sample exceeds 0.7%. At later stages of degradation, severe cracking often is accompanied by delamination and exfoliation, and may eventually lead to the total disintegration of the sample (Thorvaldson *et al.* 1927).

Over the years, various authors have relied on visual inspections to evaluate the resistance of laboratory samples to sulfate attack. For instance, Lerch (1961) and Stark (1989) used this approach to study the performance of test beams exposed to sulfate-rich soils. Their assessment was based on a numerical rating system ranging from 1.0 (indicating no evidence of degradation) to 6.0 (indicating failure).

However, it should be emphasized that the sole visual inspection of laboratory samples can be misleading. For instance, during an investigation of the influence of fly ash on the sulfate resistance of mortars, Day and Ward (1988) could observe expansions of 1% and more (accompanied by significant reductions of the mechanical properties of the samples) without any obvious signs of degradation. These results clearly emphasize the need for more rigorous methods to evaluate the performance of hydrated cement systems subjected to sulfate attack.

Numerous investigations have also clearly indicated that the volume instability of concrete under sulfate attack is influenced by a wide range of parameters. As can be seen in Figure 5.3, the kinetics of expansion is particularly sensitive to the pH of the test solution (Brown 1981). The detrimental effect of low pH has also been confirmed by Ferraris *et al.* (1997). The importance of pH will be discussed in further detail in Chapter 9.

Test results also indicate that the nature and the concentration of the sulfate solution also affect swelling. For instance, magnesium sulfate solutions are usually found to be more aggressive than sodium sulfate solutions (Gollop and Taylor 1992–1996; Thorvaldson *et al.* 1927). However, these observations should be considered with caution. As pointed out by Verbeck (1968) and Day (2000), the use of magnesium sulfate solution in a static soaking test (i.e. without any stirring of the solution) can result in the formation of a protective magnesium hydroxide coating on the surface of the test samples. As a result, ions from the attacking solution cannot penetrate into the material. The result is lower expansion than would have been observed in the absence of this protective layer.

A similar phenomenon has been reported for dense concrete fully submerged in sea water (Buenfeld and Newman 1986; Mehta 1991; Taylor 1997). The combined action of magnesium ($Mg^{2+}$) and carbonate ($HCO_3^-$) often result in the formation of a surface skin, typically consisting of a thin layer of brucite overlaid by a more slowly developing layer of aragonite. The presence of this surface skin has been found to protect initially dense concrete from further degradation.

It should however be emphasized that authors who have investigated the behavior of concrete exposed to sea water have not consistently reported the formation of this protective layer. For instance, Thomas *et al.* (1999) recently investigated the behavior of a series of laboratory concrete mixtures prepared at water–binder ratios ranging from 0.32 to 0.68. Some of these mixtures contained various amounts of fly ash. After twenty-eight days of curing, samples were placed in the tidal zone of the BRE marine exposure site on the Thames Estuary. After ten years of exposure, no evidence of a brucite layer on the surface of concrete could be detected. Examination by SEM showed that the surface layers were characterized by a decalcification of C-S-H with aragonite, magnesium silicate, and thaumasite being the primary reaction products.

Given the importance of the problem, the performance of field concrete exposed to sulfate-laden environments is well documented. Early reports date back to the beginning of the previous century (DePuy 1997; Lafuma 1927; Wig and Willams 1915; Wig *et al.* 1917). Since then, numerous comprehensive descriptions of the premature degradations of concrete structures exposed to sulfate solutions and sulfate-contaminated soils have been published.

As emphasized by Hamilton and Handegord (1968), the degradation of field concrete by sulfate attack does not usually result in the sudden failure of the structure. The detrimental action of sulfates is a progressive process of deterioration that often leads to collapse or to the necessity of demolition. Field reports indicate that the rate of deterioration can be particularly rapid and severe when concrete exposed to sulfates is continuously kept in saturated (or nearly saturated) conditions (Tuthill 1978).

One typical manifestation of the degradation of field concrete subjected to external sulfate attack is the expansion of structural elements and the subsequent development of cracks. Numerous reports of concrete slabs, placed directly on moist soils contaminated with sulfates (often called alkali soils), that have failed in buckling can be found in the literature (Figg 1999; Hamilton and Handegord 1968; Novak and Colville 1989; Price and Peterson 1968; Tuthill 1978). Although the onset of cracking in some cases is associated with soil expansion, the volume instability often appears to be the primary cause of distress.

The development of sulfate-induced cracking is not limited to slabs on grade. Sulfate attack has been clearly identified as the primary cause for the progressive degradation of mass concrete structures (Harboe 1982; Price and Peterson 1968; Reading 1982). One typical case involves the premature failure of concrete in the gate structure of a large submerged shipway in south-eastern United States (Terzaghi 1948). Various defects could be observed only two years after construction. The development of cracks at the pier surface was attributed to the abnormal expansion of concrete. The volume instability of concrete was ascribed to deleterious chemical reactions with sulfate ions originating from sea water.

As mentioned by St John *et al.* (1958), cracking is not the sole consequence of sulfate attack. Exfoliation and spalling are other frequent manifestations of the problem. These forms of degradation are often reported for slabs and foundations directly in contact with sulfate-contaminated soils (Haynes 2000; Haynes *et al.* 1996; Mehta 2000; Novak and Colville 1989; Tuthill 1978). The presence of efflorescing materials (typically sodium sulfate under the form of thenardite and mirabilite) is often reported as a precursor to this type of damage.

As discussed in Chapter 4, this type of degradation is often attributed to the crystallization of sulfate salts at the surface of concrete. According to Haynes *et al.* (1996), the mechanisms of degradation involve the penetration of sulfate solutions either by simple diffusion or by capillary suction when pore water evaporates from above-ground surfaces, the sulfate concentration becomes sufficiently high to cause crystallization. Changes in ambient temperature and relative humidity cause some salts to undergo cycles of dissolution and crystallization, which may be accompanied by volume expansion.

For some authors, this form of degradation should be distinguished from classical sulfate attack and referred to as "physical" sulfate deterioration (Haynes 2000; Haynes *et al.* 1996; Hime and Mather 1999). The relevance of this distinction has been addressed in Chapter 4 and will be further discussed in Chapter 8.

It should finally be emphasized that visual inspections of field concrete structures can be misleading, and should be considered with caution. Over the past decades, numerous authors have reported cases of badly degraded concrete (with little or any residual strength) that displayed no apparent signs of alteration (Hamilton and Handegord 1968; Harboe 1982; Price and Peterson 1968; Reading 1982). Further investigations of these structures showed that concrete had little residual mechanical strength, if any. These examples clearly emphasize the inherent limitations of visual inspections.

## 5.3   CONSEQUENCES OF SULFATE ATTACK ON THE MICROSTRUCTURE OF CONCRETE

In Chapter 4, the consequences of sulfate attack (from internal or external causes) on the microstructure of concrete have been extensively discussed. As emphasized in Chapter 4, the various types of sulfate degradation are often found to result in similar forms of distress, such as the development of gaps around some aggregate particles and the onset of microcracking. However, systematic microscopic observations of concrete samples suffering from composition-induced sulfate attack, DEF and external sulfate degradation indicate that the *evolution* of the microstructural damage may vary from one type of degradation to another.

Reports on the degradation of concrete by composition-induced and heat-induced (DEF) internal sulfate attack tend to indicate that degradation is usually rather homogeneous throughout the entire volume of concrete (Johansen *et al.* 1993; Johansen *et al.* 1995; St John *et al.* 1958). This is, for instance, the case for laboratory samples exposed to well-controlled conditions. As previously mentioned, cracking of field concrete structures affected by DEF has been reported to evolve from the external surfaces to the core of the element. However, systematic observations of degraded samples usually indicate the formation of deleterious ettringite throughout the entire volume of concrete. In addition, cracks often tend to propagate relatively quickly from the external surfaces through the core of the element.

On the contrary, most of the investigations published on the subject indicate that *external* sulfate attack usually proceeds by the inward movement of degradation "fronts" (Taylor 1997). For instance, using X-ray microanalyses, backscattered electron imaging and scanning electron microscopy, Gollop and Taylor (1992) found that laboratory samples immersed in sodium sulfate and magnesium sulfate solutions were characterized by a succession of layers (or zones) starting from the outer surface of the specimens. Each zone was found to be the result of a series of reactions between the external sulfate ions and the aluminate and calcium-bearing phases initially present within the material.

The presence of layers in laboratory samples tested for external sulfate attack was more recently confirmed by Wang (1994) who reported distribution curves for ettringite, gypsum, and portlandite in cement paste prisms (w/c = 0.4–0.6) immersed in a sodium sulfate solution (at 350 mmol/l and pH = 6) for fourteen days. These curves were obtained by layer-by-layer XRD analyses. Prior to the immersion in the solution, the samples were coated on all faces except two. During the immersion, one of the uncoated faces was exposed to a sulfate solution and the other was exposed to air. Test results clearly demonstrate that the material was damaged by the exposure to the sodium sulfate solution. As for the samples tested by Gollop and Taylor (1994), damage induced by sulfate attack was characterized as a transitional change in the phase distribution at the vicinity of the surface exposed to the sulfate solution (see Figure 5.4).

It should be emphasized that the formation of layers upon sulfate attack is not solely limited to laboratory specimens. Systematic observations performed by various authors also indicate the presence of reaction zones in field concrete samples. For instance, examination by St John (1982) of thin sections of concrete taken from tunnel sections exposed to ground water contaminated with sodium sulfate (and containing less than 50 ppm of $SO_4$) revealed the presence of exfoliated layer near the surface. The altered layer was characterized by the presence of cracks filled with gypsum, which graded abruptly into apparently sound concrete.

More recently, Diamond and Lee (1999), Ju *et al.* (1999), Brown and Doerr (2000) and Brown and Badger (2000) made similar observations for samples of permeable concrete originating from flatworks which had been exposed

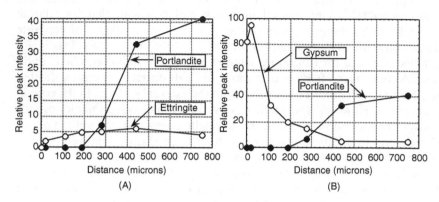

*Figure 5.4* Phase distributions for moderate sulfate resistant cement, w/c = 0.6, cured
for seven days, and immersed at pH = 6 for fourteen days.
*Source*: Wang (1994)

for many years to sulfate-bearing soils. Extensive examinations of these samples
using backscatter mode SEM and Energy-Dispersive X-Ray Analyses (EDXA)
revealed the presence of deposits of crystalline sulfates on the upper surfaces
of the slabs. Zones of degradation and reaction fronts could also be discerned
in the bottom portion of the slabs in contact with the soil. In many cases, the
analyses revealed the presence of gypsum at the vicinity of the lower surface of
the cores. A second zone with extensive ettringite formation was routinely found
to lie above the layer of gypsum. Although the precipitation of sulfate-bearing
phases in these porous systems had apparently not resulted in significant macroscopic expansion, it had clearly contributed to the formation of microcracks.

As will be seen in the following section, the formation of these layers readily
complicates the study of the detrimental influence sulfate attack on the
engineering properties of concrete. Given the heterogeneous nature of the
degradation process, it is usually difficult to isolate the effect of a single
phenomenon (such as ettringite formation).

## 5.4   CONSEQUENCES OF SULFATE ATTACK ON THE
MECHANICAL PROPERTIES OF CONCRETE

Reports on the consequences of composition-induced internal sulfate attack
clearly indicate that this form of degradation not only results in the formation of a network of microscopic and macroscopic cracks but also contributes
to significantly reduce the mechanical properties of concrete.

The effect of composition-induced sulfate attack on the volume stability
and compressive strength of a series of mortars is presented in Figure 5.5.
These results reported by Ouyang *et al.* (1988) were obtained for mixtures to
which excess gypsum (added under the form of phosphogypsum) was added.

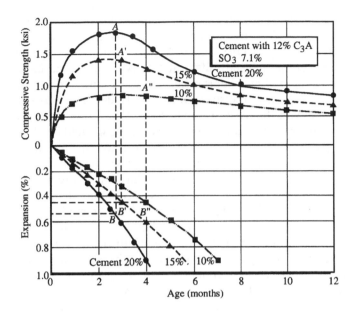

*Figure 5.5* Effect of cement content on strength and expansion of mortars under internal sulfate attack ($1.0\,ksi \times 6.89\,MPa$).
*Source*: Ouyang *et al.* (1988)

As can be seen, the three compressive strength curves show exactly the same behavior. Despite the fact that all mixtures were found to expand right from the beginning of the tests, the compressive strength of all samples initially increased. This initial gain is probably due to continuous hydration and eventually to the pore filling effect of ettringite and gypsum formation. After a few months of curing, compressive strength values were found to peak and then decrease. Significant reductions were observed for the mixtures prepared at a cement content of 20% and 15% respectively.

The authors found that that the cement content of the mixture influenced the admissible expansion beyond which the compressive strength of mortar was detrimentally affected. They proposed the following relationship to predict the value of this critical expansion (noted $E$ and expressed in mm/mm):

$$E = 0.0026 + 0.01C \tag{5.1}$$

where $C$ is the cement content (expressed as a percentage of the total mass of the mixture).

The results reported by Ouyang *et al.* (1988) are in good agreement with observations of field concrete affected by composition-induced internal sulfate attack (Harboe 1982; St John *et al.* 1958). As previously mentioned, degradation is usually found to result in severe cracking and significant strength reduction.

Studies on the consequences of DEF also indicate that this form of internal sulfate attack can also have detrimental effects on the compressive strength of concrete. Data reported by Lewis *et al.* (1995) indicate that heat-cured mortars affected by DEF tend to behave similarly to samples suffering from composition-induced sulfate attack. After an initial increase, the compressive strength of most mortar samples was found to decrease significantly. The time at which the strength started to decrease was believed to correspond well with the onset of expansion. However, no systematic relationship between expansion and compressive strength could be observed.

Over the past decade, numerous authors have investigated the influence of external sulfate attack on the mechanical properties of concrete. Most studies published on the subject tend to indicate that the reorganization of the internal microstructure of concrete is accompanied by a significant reduction of the material's strength and rigidity (elastic modulus).

As previously mentioned, the layered damage resulting from external sulfate attack has often complicated the work of researchers. In order to work on a relatively homogeneous material (for which degradation was more or less uniform throughout its volume), many authors have elected to test relatively small samples. The reduced dimensions of the samples also explain why many studies were conducted on mortar samples. In that respect, one should be cautious in interpreting and comparing such tests vis-à-vis samples of concrete of larger dimensions.

Typical laboratory results on the effect of sulfate attack on the compressive and tensile strengths of mortar mixtures are given in Figures 5.6 and 5.7. As can be seen, the immersion in a sulfate solution initially results in a slight increase in strength (Brown 1981; Jambor 1998; Thomas *et al.* 1999; Thorvaldson *et al.* 1927). A similar trend was also reported for the stress corrosion of well-cured paste mixtures subjected to flexural loading while immersed in a $Na_2SO_4$ solution (Werner *et al.* 2000). After this initial gain, the material usually loses strength very rapidly.

As discussed in Chapter 4, the initial gain in strength might be associated to hydration effects and to the precipitation of ettringite, which tends to reduce the porosity of the material. The subsequent loss in strength and rigidity of the material is often attributed to expansion and the onset of microcracking. This analysis is in good agreement with the data of Ouyang *et al.* (1988) and those of Ouyang (1989) which show that the beginning of the strength loss usually corresponds to an expansion of 0.1% (see Figure 5.8).

Results on the initial strength gains of laboratory samples subjected to external sulfate attack should be considered with caution. As can be seen in Figure 5.6, no such increase can be observed for samples immersed in low pH solutions. This phenomenon can probably be explained by the fact that low pH solutions favor the formation of ettringite and the dissolution of calcium hydroxide crystals (Marchand *et al.* 1999a). Over the past decades, numerous studies have clearly emphasized the detrimental influence of portlandite

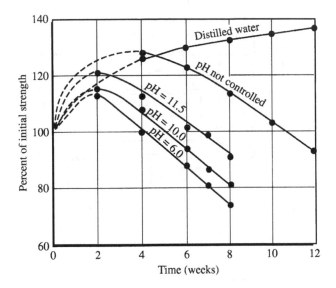

*Figure 5.6* Variation in cube strength with time under the following experimental conditions: distilled water immersion, immersion in 0.35 M Na₂SO₄ without pH control, and immersion in 0.35 M Na₂SO₄ while maintaining the solution pH at 6, 10 and 11.5.

*Source*: Brown (1981)

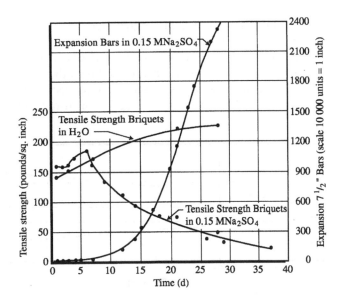

*Figure 5.7* Comparison of expansion of mortar bars and tensile strength of bricquets in 0.15 M solution of Na₂SO₄ at 22 °C.

*Source*: Thorvaldson *et al.* (1927)

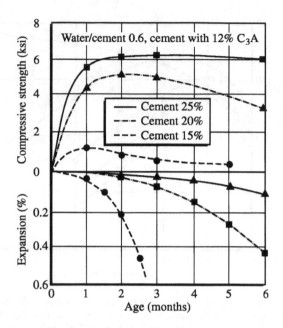

*Figure 5.8* Effect of cement content on strength and expansion of mortars under external sulfate attack (1.0 ksi = 6.89 MPa).
*Source*: Ouyang *et al.* (1988)

dissolution on the mechanical properties of hydrated cement systems (Marchand *et al.* 1999b; Terzaghi 1948; Tremper 1931; Saito and Deguchi 2000).

The reorganization of the internal microstructure of concrete subjected to sulfate attack also tends to increase the susceptibility of the material to stress corrosion. Typical results obtained by Schneider and Piasta (1991) are given in Figure 5.9. As can be seen, the combined action of sustained flexural load

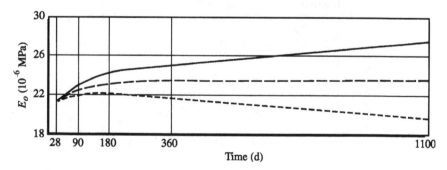

*Figure 5.9* Change in the elastic modulus of concrete under load immersed in various solutions.
*Source*: Schneider and Piasta (1991)

and chemical attack by a $Na_2SO_4$ solution clearly contributes to accelerate the reduction in the elastic modulus of the material.

Reports on the behavior of field concrete tend to confirm the detrimental effect of external sulfate attack on the mechanical properties of the material. For instance, many authors reported marked losses in compressive strength for concrete structures exposed to sulfate-laden soils or sea water (Hamilton and Handegord 1968; Harboe 1982; Price and Peterson 1968; Terzaghi 1948).

Curiously, it appears that the shear resistance and tensile strength of concrete are more sensitive to external sulfate attack than compressive strength (Hamilton and Handegord 1968; Harboe 1982). This phenomenon can probably be explained (at least in part) by the layered damage resulting from external sulfate attack. The tensile strength and the shear resistance of concrete are more sensitive by internal defects than compressive strength.

## 5.5   CONCLUDING REMARKS

As can be seen, whatever its form, sulfate attack has significant consequences on the microstructure and engineering properties of concrete. Marked expansion and loss in the mechanical properties of the material often accompany sulfate-induced microstructural alterations. However, as emphasized by Skalny and Pierce (1999), damage induced by sulfate can be easily avoided (or at least limited) by simple measures. This point will be briefly reviewed in the following chapter.

## REFERENCES

Brown, P.W. (1981) "An evaluation of the sulfate resistance of cements in a controlled environment", *Cement and Concrete Research* **11**: 719–727.

Brown, P.W. and Badger, S. (2000) "The distributions of bound sulfates and chlorides in concrete subjected to mixed NaCl, $MgSO_4$, $Na_2SO_4$ attack", *Cement and Concrete Research* **30**: 1535–1542.

Brown, P.W. and Doerr, A. (2000) "Chemical changes in concrete due to ingress of chemical species", *Cement and Concrete Research* **30**: 411–418.

Buenfeld, N.R. and Newman, J.B. (1986) "The resistivity of mortars immersed in sea water", *Cement and Concrete Research* **16**: 511–524.

Crammond, N.J. (1984) "Examination of mortar bars containing varying percentages of coarsely crystalline gypsum as aggregate", *Cement and Concrete Research* **14**: 225–230.

Day, R.L. (1992) "The effect of secondary ettringite formation on the durability of concrete: a literature analysis", *PCA R&D Bulletin RD108T*.

Day, R.L. (2000) "Development of performance tests for sulfate attack on cementitious systems", *Cement Concrete and Aggregates* **22**: 169–178.

Day, R.L. and Ward, M.A. (1988) "Sulphate durability of plain and fly ash mortars", *MRS Symposia Proceedings*, **113**: 153–161.

DePuy, G.W. (1994) "Chemical resistance of concrete", in Lamond and Klieger (eds) *Tests and Properties of Concrete*, STP 169C, ASTM, Philadelphia, 263.

DePuy, G.W. (1997) "Review of sulfate attack in US and Canada", report submitted on behalf of plaintiffs in *Murphy v. First Southwest Diversified Partners*, Case #752593 (Orange County Superior Court, California, 1997), November 3, 89 pp.

Diamond, S. (1996) "Delayed ettringite formation – processes and problems", *Cement and Concrete Composites* **18**: 205–215.

Diamond, S. and Lee, R.J. (1999) "Microstructural alterations associated with sulfate attack in permeable concretes", in J. Marchand and J. Skalny (eds) *Materials Science of Concrete Special Volume: Sulfate Attack Mechanisms*, The American Ceramic Society, Westerville, OH, pp. 123–173.

Famy, C. (1999) "Expansion of heat-cured mortars", Ph.D thesis, University of London, September.

Ferraris, C.F., Clifton, J.R., Stutzman, P.E. and Gaboczi, E.J. (1997) "Mechanism of degradation of Portland cement-based systems by sulfate attack" in K. Scrivener and J.F. Young (eds) *Mechanism of Chemical Degradation of Cement-based Systems*, E & FN Spon, pp. 185–172.

Figg, J. (1999) "Field studies of sulfate attack on concrete", in J. Marchand and J. Skalny (eds) *Materials Science of Concrete Special Volume: Sulfate Attack Mechanisms*, The American Ceramic Society, Westerville, OH, pp. 315–324.

Gollop, R.S. and Taylor, H.F.W. (1992–1996) "Microstructural and microanalytical studies of sulfate attack I. Ordinary cement paste. II. Sulfate resisting Portland cement: ferrite composition and hydration chemistry. III. Sulfate-resisting Portland cement: reactions with sodium and magnesium sulfate solutions. IV. Reactions of a slag cement paste with sodium and magnesium sulfate solutions. V. comparison of different slag blends", *Cement and Concrete Research* 22: 1027–1038; 24: 1347–1358; 25: 1581–1590; 26: 1013–1028; 26: 1029–1044.

Hamilton, J.J. and Handegord, G.O. (1968) "The performance of ordinary Portland cement concrete in prairies soils of high sulphate content", in E.G. Swenson (ed.) *Performance of Concrete-Resistance of Concrete to Sulphate and Other Environmental Conditions: A Symposium in Honour of Thobergur Thorvaldson*, University of Toronto Press, pp. 135–158.

Harboe, E.M. (1982) "Longtime studies and field experiences with sulfate attack", in *Sulfate Resistance of Concrete* (George Verbeck Symposium), ACI SP-77, pp. 1–20.

Haynes, H. (2000) "Sulfate attack on concrete: laboratory versus field experience", *Suppl. Proc. 5th CANMET/ACI Int.Conf. Durability of Concrete*, Barcelona, June (in press).

Haynes, H., O'Neill, R. and Mehta, P.K. (1996) "Concrete deterioration from physical attack by salts", *Concrete International* 18(1) (January): 63–68.

Hime, W.G. and Mather, B. (1999) "'Sulfate Attack' or is it?", *Cement and Concrete Research*, **29**: 789–791.

Jambor, J. (1998) "Sulfate corrosion of concrete", unpublished manuscript summarizing Dr Jambor's work on the *Sulfate durability of concrete*. (The author passed away in May 1998.)

Johansen, V., Thaulow, N. and Skalny, J. (1993) "Simultaneous presence of alkali-silica gel and ettringite in concrete", *Adv. Cem. Res.* **5**: 23–29.

Johansen, V., Thaulow, N. and Skalny, J. (1995) "Internal reactions causing cracking of concrete", *Beton + Fertigteil Technik* **11**: 56–68.

Ju, J.W., Weng, L.S., Mindess, S. and Boyd, A.J. (1999) "Damage assessment and service-life prediction of concrete subject to sulfate attack", in J. Marchand and

J. Skalny (eds) *Materials Science of Concrete Special Volume: Sulfate Attack Mechanisms*, The American Ceramic Society, Westerville, OH, pp. 265–282.

Lafuma H. (1927) "Theory of expansion of cement" (in French), *Revue des Materiaux de Construction et de Traveaux Public* **243**: 441–444.

Lagerblad, B. (1999) "Long-term test of concrete resistance against sulfate attack", in J. Marchand and J. Skalny (eds) *Materials Science of Concrete Special Volume: Sulfate Attack Mechanisms*, The American Ceramic Society, Westerville, OH, pp. 325–336.

Lawrence C.D. (1995a) "Delayed ettringite formation: an issue?", in J. Skalny and S. Mindess (eds) *Material Science of Concrete*, vol. IV, The American Ceramic Society, Westerville, OH, 113 pp.

Lawrence, C.D. (1995b) "Mortar expansions due to delayed ettringite formation", *Cem. Concr. Res.* **25**: 903–914.

Lerch, W. (1961) "Significance of tests for sulfate resistance", *Proceedings, ASTM STP* **61**: 1043–1054.

Lewis, M.C., Scrivener, K.L. and Kelham, S. (1995) "Heat curing and delayed ettringite formation", *MRS Symposia Proceedings* **370**: 67–76.

Mather, K. (1982) "Current research in sulfate resistance at the Waterway Experiment Station", ACI SP-77, pp. 63–74.

Marchand, J., Samson, E. and Maltais, Y. (1999a) "Modeling microstructural alterations of concrete subjected to sulfate attack", in J. Marchand and J. Skalny (eds) *Materials Science of Concrete Special Volume: Sulfate Attack Mechanisms*, The American Ceramic Society, Westerville, OH, pp. 211–257.

Marchand, J., Beaudoin, J.J. and Pigeon, M. (1999b) "Influence of calcium hydroxide dissolution on the engineering properties of cement-based materials", in J. Marchand and J. Skalny (eds) *Materials Science of Concrete Special Volume: Sulfate Attack Mechanisms*, The American Ceramic Society, Westerville, OH, pp. 283–294.

Mehta, P.K. (1991) *Concrete in the Marine Environment*, Elsevier Applied Science, London.

Mehta, P.K. (1992) "Sulfate attack on concrete – a critical review", in J. Skalny (ed.) *Material Science of Concrete*, American Ceramic Society, Westerville, OH, pp. 105–130.

Mehta, P.K. (2000) "Sulfate attack on concrete: separating the myth from reality", *Concrete International* **22**(8): 57–61.

Novak, G.A. and Colville, A.A. (1989) "Effloresence mineral assemblages associated with cracked and degraded residential concrete foundation in southern California", *Cem. Concr. Res.* **19**(1): 1–6.

Odler, I. and Jawed, I. (1991) "Expansive reactions in concrete" in J. Skalny and S. Mindess (eds) *Materials Science of Concrete*, The American Ceramic Society, Westerville OH, pp. 221–247.

Ouyang, C. (1989) "A damage model for sulfate attack of cement mortars", *Cement, Concrete, and Aggregates*, CCAGDP, **11**(2): 92–99.

Ouyang, C., Naani, A. and Chang, W.F. (1988) "Internal and external sources of sulfate ions in Portland cement mortar: two types of chemical attack", *Cement and Concrete Research* **18**: 699–709.

Patzias, T (1987) "An evaluation of sulfate resistance of hydraulic-cement mortars by the ASTM C 1012 test method", ACI SP-100, pp. 2101–2120.

Price, C.G. and Peterson, R. (1968) "Experience with concrete in sulphate environments in Western Canada", in E.G. Swenson (ed.) *Performance of Concrete-Resistance of Concrete to Sulphate and Other Environmental Conditions: A Symposium in Honour of Thobergur Thorvaldson*, University of Toronto Press, pp. 93–112.

Reading, T.J. (1982) "Physical aspects of sodium sulfate attack on concrete", ACI SP-77, pp. 75–81.

Saito, H. and Deguchi, A. (2000) "Leaching tests on different mortars using accelerated electrochemical method", *Cement and Concrete Research* **30**: 1815–1825.

Samarai, M.A. (1976) "The disintegration of concrete containing sulphate-contaminated aggregates", *Magazine of Concrete Research* **28**: 130–142.

Schneider, U. and Piasta, W.G. (1991) "The behaviour of concrete under Na$_2$SO$_4$ solution attack and sustained compression or bending", *Magazine of Concrete Research* **43**: 281–289.

Skalny, J. and Pierce, J. (1999) "Sulfate attack: an overview", in J. Marchand and J. Skalny (eds) *Materials Science of Concrete Special Volume: Sulfate Attack Mechanisms*, The American Ceramic Society, Westerville, OH, pp. 49–63.

Smith, F.L. (1958) "Effect of calcium chloride addition on sulfate resistance of concrete placed and initially cured at 40 and 70 °F", *Concrete Laboratory Report No. C-900*, Bureau of Reclamation, Denver, CO.

Stark, D. (1989) "Durability of concrete in sulfate-rich soils", in *Research and Development Bulletin RD097.01T*, Portland Cement Association, Stokie, Illinois.

St John, D.A. (1982) "An unsual case of ground water sulfate attack on concrete", *Cement and Concrete Research* **12**: 633–639.

St John, D.A., Poole, A.W. and Sims, I. (1958) *Concrete Petrography – A Handbook of Investigative Techniques*, Arnold Publisher, London.

Taylor H.W.F. (1997) *Cement Chemistry*, 2nd edn, Thomas Telford, London.

Terzaghi, R.D. (1948) "Concrete deterioration in a shipway", *Journal of the American Concrete Institute* **44**: 977–1005.

Thomas, M.D.A., Bleszynski, R.F. and Scott, C.E. (1999) "Sulphate attack in marine environment", in J. Marchand and J. Skalny (eds) *Materials Science of Concrete Special Volume: Sulfate Attack Mechanisms*, The American Ceramic Society, Westerville, OH, pp. 301–313.

Thorvaldson, T. (1952) "Chemical aspects of the durability of cement products", *Proc. 3rd International Symposium on the Chemistry of Cements*, London, UK, pp. 436–484.

Thorvaldson, T., Harris, R.H. and Wolochov, D. (1925) "Disintegration of Portland cement in sulfate waters", *Industrial and Engineering Chemistry* **17**(3): 467–470.

Thorvaldson, T., Lamour, R.K. and Vigfusson, V.A. (1928) "The expansion of Portland cement mortar bars during disintegration in sulphate solutions", *The Engineering Journal* **10**(4): 199–206.

Thorvaldson, T., Vigfusson, V.A. and Larmour, R.K. (1927) "The action of sulfates on the components of Portland cement", *Trans. Royal. Soc. Canada*, 3rd Series, **21**, Section III, 295.

Tremper B. (1931) "The effect of acid waters on concrete", *Journal of The American Concrete Institute-Proceedings* **28**: 1–32.

Tuthill, L.H. (1936) "Resistance of cement to the corrosive action of sulphate solutions", *Journal of The American Concrete Institute-Proceedings* **88**: 83–106.

Tuthill, L.H. (1978) "*Resistance to chemical attack*", *ASTM STP-169-A*, pp. 369–387.

Verbeck, G.J. (1968) "Field and laboratory studies of the sulphate resistance of concrete", in E.G. Swenson (ed.) *Performance of Concrete-Resistance of Concrete to Sulphate and Other Environmental Conditions: A Symposium in Honour of Thobergur Thorvaldson*, University of Toronto Press, pp. 113–125.

Wang, J.G. (1994) "Sulfate attack on hardened cement paste", *Cement and Concrete Research* **24**: 735–742.

Werner, K.C., Chen, Y. and Odler, I. (2000) "Investigations on stress corrosion of hardened cement pastes", *Cement and Concrete Research* **30**: 1443–1451.

Wig, R.J. and Williams, G.M. (1915) "Investigations on the durability of cement drain tiles in alkali soils", *Technological Papers of the Bureau of Standards*, no. 44, Washington, DC: US Government Printing Office, 56 pp.

Wig, R.J., Williams, G.M. and Finn, A.N. (1917) "Durability of cement drain tiles and concrete in alkali soils", *Technological Papers of the Bureau of Standards*, no. 95, Washington, DC: US Government Printing Office, 95 pp.

# 6 Prevention of sulfate attack

## 6.1 INTRODUCTION

Concrete in service is often exposed to aggressive environments. Although severe exposure conditions may sometime be at the origin the premature degradation of concrete, durability problems often originate from an improper production and use of the material. As mentioned in the first chapter of this monograph, man abuses concrete in various ways, most of them based on an insufficient knowledge of the material.

It should however be emphasized that it is relatively simple and economical to produce durable concrete. We have numerous examples of durable concrete structures that have performed as expected for decades while being exposed to severe conditions. In all cases, concrete had been produced and handled with care.

As discussed in Chapters 4 and 5, the widespread occurrence and destructiveness of sulfate attack led to many investigations over the years into the mechanism of deterioration. Many of these studies have also allowed identification of practical solutions to protect concrete against sulfate attack. These prevention methods are briefly reviewed in the following paragraphs.

## 6.2 MEASURES TO PROTECT CONCRETE AGAINST COMPOSITION-INDUCED INTERNAL SULFATE ATTACK

As previously mentioned in Chapter 4, cement itself may be a source of excessive sulfate in concrete. This is the reason why requirements of CEN, ASTM (see ASTM C150; ASTM C1157, BS 5328) and other standards on cement and clinker composition should be closely followed; this will assure proper *concentrations and ratios of the relevant clinker minerals* to give sulfate levels that will not lead to excessive expansion.

Aggregates and mineral additives are other potential sources of excessive sulfate. Selected *aggregates and intermixed mineral admixtures should not contain sulfate-bearing compounds* that may later become available for reaction with cement components of the concrete mixture.

Quality Control is clearly one key issue in the protection of concrete against composition-induced internal sulfate attack. It is therefore recommended to *continuously monitor* not only the sulfate content of the ground and shipped cement but, on a regular basis, also the content and form of sulfates present in the cement clinker and in the inter-ground supplementary materials. Determination of optimum gypsum content should become a routine test on a schedule more frequent than is the case in the majority of cement operations at the present time. Aggregate and admixtures should be analyzed for presence of sulfates.

It is most important to *maintain proper records* of the analytical and mechanical tests, and to make them available to the customers upon request.

## 6.3 MEASURES TO PROTECT CONCRETE AGAINST HEAT-INDUCED INTERNAL SULFATE ATTACK

Proper mixture design is one way of protecting concrete against degradation by heat-induced internal sulfate attack. The materials used in designing concrete mixtures (cement, aggregate, supplementary materials, admixtures) must *pass the existing specifications and have proven history of satisfactory performance. The lowest possible w/cm is recommended.* Under proper conditions, to be defined below, there is no evidence showing that regular Portland or blended cements, and most aggregates that passed the required specifications, would cause unexpected durability problems related to heat curing. Use of some, but not all, mineral admixtures may be beneficial; prior testing is recommended.

Obviously, special care should be taken during the casting and curing operations. The formwork material and its thickness affect the heat transfer, and this fact must be taken into consideration when designing for homogeneous heat and humidity distribution within the concrete member. Exposed *concrete surfaces should be kept wet*, but condensed water should not drip on them. When concrete members are stacked within the curing chamber, *even distribution of heat and humidity should be maintained by proper circulation.*

*Precuring or preset time must be adequate* to allow the cement used in making the concrete to set properly. Depending on the type of cement used, this may be between two and four hours. It should be kept in mind that prematurely heated fresh concrete will develop lower strength and may lead to decreased durability.

*Heating rate of concrete should be steady* (limited to about 15–20 °C, or about 25–35 °F, maximum per hour) and the *temperature rise uniformly distributed for the whole member as well as within the curing chamber*; such curing procedure will prevent formation of microstructural faults and cracks that may adversely affect long-term durability of the treated concrete member. *Heat treatment must not lead to drying out* of exposed surfaces, as drying and heating may result in pore coarsening; this is best achieved by maintaining adequately high relative humidity and its homogeneous distribution within the curing chamber.

In designing concrete to be exposed to heat treatment, *the heat of hydration of the cement should always be taken into consideration*: it is the total heat input (heat of the ambient temperature plus heat of hydration plus heat added during curing) that controls the quality of the product! In our opinion, it is prudent to *limit the maximum concrete temperature to below 65 °C or about 150 °F*. Because heat dissipation is an important aspect of curing, the size of the concrete member has to be taken into account.

Measures must be established allowing controlled cooling and prevention of premature dry-out; such measures allow elimination or reduction of crack formation and lead to improved durability by decreasing the permeability. *The difference between the external and maximum internal temperature of a member should never exceed 20 °C or about 35 °F.*

Quality control by concrete producers is a must! The most important aspect of proper heat-cured concrete making is the *control of the time-temperature regime*. The temperature and its proper (homogeneous) distribution in the curing chamber should be monitored continuously and, more important, *a good correlation between the curing chamber temperature and the temperature of the concrete* should be maintained.

## 6.4  MEASURES TO PROTECT CONCRETE AGAINST EXTERNAL SULFATE ATTACK

One of the primary conditions for the proper design and erection of a concrete structure is the full understanding of all aspects of the local environment. Of special importance for sulfate resistance is *detailed knowledge of the soil and ground water conditions*, including the presence, homogeneity of distribution, chemistry, and mineralogy of the sulfate-containing species. This is important for more detailed understanding of the chemical interactions that may occur between the sulfates and the concrete components. Of equal importance is *understanding of the ground water movement and depth*. We suggest that, *if the sulfate concentration at the job site is variable, concrete (particularly the w/cm, cement type and content) should be specified for the highest observed sulfate level.*

Understanding of the ranges and fluctuations in temperature and humidity enables proper selection of concrete quality needed in the given environment. Therefore, these *environmental/atmospheric conditions should be taken into consideration right at the design stage*, to assure proper concrete mix design and structural design, minimizing the access of ground water to the structure. Understandably, good workmanship during the structure erection is crucial. It is important to remember that the potential negative effect of atmospheric conditions can be completely negated if concrete quality compatible with the environment is delivered.

The three main strategies for improving resistance to sulfate solutions are: (1) making a high quality, impermeable concrete; (2) using a sulfate-resistant binder; and (3) making sure that concrete is properly placed and cured on site.

As specified by many standards and codes (see ACI and UBC), concrete should be designed to give dense, low-porosity concrete matrix that can resist penetration of aggressive chemical species into hardened concrete. Depending on the sulfate concentration in the environment of use, *maximum w/cm of 0.5, but possibly as low as 0.4, is recommended* (see for instance Table 1.5). It should be also taken into consideration that achievement of the specified minimum compressive strength may not be an adequate measure of durability under the given environmental conditions. Therefore, an *increase in cement content above that needed to achieve the minimum strength* (*while keeping the w/cm low!*) *should be considered*.

As discussed in Section 4.10, ASTM Type V and other "sulfate-resisting" cements were specifically developed for use in sulfate-rich environments. *They should be used with the understanding that they are not a panacea against sulfate attack unless used with quality concrete. Sulfate resisting cements are not a substitute for proper concrete making.* Their use is recommended to give a secondary level of protection *in addition* to (not instead of!) protection given by low water–cement ratio, adequate cement content, and overall proper mix design and good workmanship. It should be borne in mind that in cases where the aggressive sulfates are present as magnesium sulfates, Type V and similar cements may not give the desired protection.

In addition, in severe sulfate environments, *the use of appropriate and tested mineral admixtures may be desirable*. However, special care should be taken to select the proper source of fly ash and/or slag. Information on the influence of these two types of mineral additives indicates that the influence of these materials on the performance of concrete tends to vary significantly from one source to another (Mehta 1986; Stark 1989).

Control of concrete quality and workmanship is most desirable. *The needed knowledge and technologies are mostly available; their proper use must be expanded*. In well-designed concrete and concrete structures, and under proper management of the concrete processing and structure erection, the danger of sulfate attack can be completely eliminated.

Many engineers are often tempted to rely on impermeable barriers to prevent sulfate solutions from coming into contact with the concrete. As emphasized by DePuy (1994), impermeable barriers are not recommended as a long-term solution to an aggressive sulfate solution as there is no guarantee that the barrier will perform effectively over the required period.

## 6.5 CONCLUDING REMARKS

As can be seen, practical solutions to protect concrete against sulfate attack are, in most cases, simple and economical. Most of these solutions rely on a good understanding of the material and the exposure conditions. It should also be stressed that the costs related to the repair or the partial reconstruction of a concrete structure affected by sulfate attack are usually much more important than those required to prevent the problem.

## REFERENCES

ACI (1992) ACI 201.2R-92, *Guide to Durable Concrete*, ACI.

ACI 318-99 (1999) "Building code requirements for structural concrete", American Concrete Institute, Farming Hill, MI.

ASTM C150-95 (1995) ASTM Standard Specification for Cement, C 150–195

ASTM C1157M "Standard performance specification for blended hydraulic cement" ASTM, Philadelphia, PA.

BS 5328 (1997) "British Standard 5328: Concrete Part 1", *Guide to specifying concrete*, British Standard Institution, Issue 2, May 1999.

CEN (1998) CEN/TC 104/SC 1 N 308, "Common rules for precast concrete products" (draft 04/98), September.

DePuy, G.W. (1994) "Chemical resistance of concrete," in Lamond and Klieger (eds) *Tests and Properties of Concrete*, STP 169C, ASTM, Philadelphia, 263.

Mehta, P.K. (1986) "Effect of fly ash composition on sulfate resistance of cement", *ACI Materials Journal* 83: 994–1000.

Stark, D. (1989) "Durability of concrete in sulfate-rich soils", in *Research and Development Bulletin RD097.01T*, Portland Cement Association, Stokie, Illinois.

Uniform Building Code (1997) "Concrete", Chapter 19, in *Structural Engineering Design Provisions*, vol. 3, International Conference of Building Officials.

# 7 Modeling of deterioration processes

## 7.1 INTRODUCTION

Hydrated cement systems are used in the construction of a wide range of structures. During their service life, many of these structures are exposed to various types of chemical aggression involving sulfate ions. In most cases, the deterioration mechanisms involve the transport of fluids and/or dissolved chemical species within the pore structure of the material. This transport of matter (in saturated or unsaturated media) can either be due to a concentration gradient (diffusion), a pressure gradient (permeation), or capillary suction. In many cases, the durability of the material is controlled by its ability to act as a tight barrier that can effectively impede, or at least slow down the transport process.

Given their direct influence on durability, mass transport processes have been the objects of a great deal of interest by researchers. Although the existing knowledge of the parameters affecting the mass transport properties of cement-based materials is far from being complete, the research done on the subject has greatly contributed to improve the understanding of these phenomena. A survey of the numerous technical and scientific reports published on the subject over the past decades is beyond the scope of this report, and comprehensive reviews can be found elsewhere (Nilsson *et al.* 1996; Marchand *et al.* 1999).

As will be discussed in the last chapter of this book, the assessment of the resistance of concrete to sulfate attack by laboratory or *in situ* tests is often difficult and generally time-consuming (Harboe 1982; Clifton *et al.* 1999; Figg 1999). For this reason, a great deal of effort has been made towards developing microstructure-based models that can reliably predict the behavior of hydrated cement systems subjected to sulfate attack.

A critical review of the most pertinent models proposed in the literature is presented in this chapter. Some of these models have been previously reviewed by other authors (Clifton 1991; Clifton and Pommersheim 1994; Reinhardt 1996; Walton *et al.* 1990). The purpose of this chapter is evidently not to duplicate the works done by others, but rather to complement them.

In the present survey, emphasis is therefore placed on the most recent developments on the subject. Empirical, mechanistic and numerical models are reviewed in separate sections. Special attention is paid to the recent innovations in the field of numerical modeling. Recent developments in computer engineering have largely contributed to improve the ability of scientists to model complex problems (Garboczi 2000). As will be seen in the last section of this chapter, numerous authors have taken advantage of these improvements to develop new models specifically devoted to the description of the behavior of hydrated cement systems subjected to chemical attack.

It should be emphasized that this review is strictly limited to microstructure-based models developed to predict the performance of concrete subjected to sulfate attack. Over the years, some authors have elaborated various kinds of empirical equations to describe, for instance, the relationship between sulfate-induced expansion to variation in the dynamic modulus of elasticity of concrete (Smith 1958; Biczok 1967). These models are not discussed in this chapter.

It should also be mentioned that this chapter is exclusively restricted to models devoted to the behavior of concrete subjected to *external* sulfate attack. Despite the abundant scientific and technical literature published on the topic over the past decade, the degradation of concrete by internal sulfate attack has been the subject of very little modeling work.

## 7.2   MICROSTRUCTURE-BASED PERFORMANCE MODELS

Over the past decades, authors have followed various paths to develop micro-structure-based models to predict the behavior of hydrated cement systems subjected to sulfate attack. Models derived from these various approaches may be divided into three categories: empirical models, mechanistic (or pheno-menological) models, and computer-based models. Although the limits between these categories are somewhat ambiguous, and the assignment of a particular model in either of these classes is often arbitrary, such a classifica-tion has proven to be extremely helpful in the elaboration of this chapter. It is also believed that this classification will contribute to assist the reader in evaluating the limitations and the advantages of each model.

Before reviewing the various models found in the literature, the characteris-tics of a good model deserve to be defined. The main quality of such a model lies in its ability to reliably predict the behavior of a wide range of materials. As mentioned by Garboczi (1990), the ideal model should also be based on direct measurements of the pore structure of a representative sample of the material. These measurements should be of microstructural parameters that have a direct bearing on the durability of the material, and the various char-acteristics of the porous solid (e.g. the random connectivity and the tortuosity of the pore structure, the distribution of the various chemical phases...) should be treated realistically. As can be seen, the difficulties of developing a good model are as much related to the identification and the measurement of relevant microstructural parameters than to the subsequent treatment of this information.

### 7.2.1 Empirical models

As emphasized by Kurtis *et al.* (2000), concrete mixtures are typically designed to perform for 50–100 years with minimal maintenance. However, the premature degradation of numerous structures exposed to sea water and sulfate soils has raised many questions with respect to the long-term durability of concrete under chemically aggressive conditions. As reviewed in Chapter 4, these concerns have motivated many researchers to investigate the mechanisms of external sulfate attack.

Engineers have also tried to develop various approaches to estimate the long-term durability of concrete structures subjected to sulfate attack. Early attempts to predict the remaining service life of concrete were relatively simple and mainly consisted in linear extrapolations based on a given set of experimental data (Kalousek *et al.* 1972; Terzaghi 1948; Verbeck 1968).

Following these initial efforts, many authors have later tried to elaborate more sophisticated ways to predict the durability of concrete. Most of these early service-life models essentially consist in empirical equations. All of them have been developed using the same approach. An equation linking the behavior of the material to its microstructural properties is deduced from a certain number of experimental data. In most cases, the mathematical relationship is derived from a (more or less refined) statistical analysis of the experimental results.

Jambor (1998) is among the first researchers to develop an empirical equation describing the rate of "corrosion" of hydrated cement systems exposed to sulfate solutions. The equation is derived from the analysis of a large number of experimental data obtained over a fifteen-year period. The objective of this comprehensive research program was to investigate the behavior of 0.6 water–binder ratio mortar mixtures totally immersed in sodium sulfate ($Na_2SO_4$) solutions.

During the course of Dr Jambor's project, eight different Portland cements were tested. The $C_3A$ content of these cements ranged from 9 to 13% (as calculated according to Bogue's method). Nine additional mixtures were prepared with a series of four granulated blast-furnace slag binders (with a slag content ranging from 10 to 70%) and another series of five blended cements containing 10, 20, 30, 40, and 50% of volcanic tuff as a pozzolanic admixture. All the blended mixtures were prepared in the laboratory with the Portland cement made of 11.5% $C_3A$.

All mixtures were moist cured during twenty-eight days and then immersed in the sodium sulfate solutions. The test solutions were prepared at various concentrations ranging from 500 to 33,800 g/l of $SO_4$. During the entire course of the project, the sulfate solution to sample volume ratio was kept constant at ten and the test solutions were systematically renewed in order to maintain the sulfate concentration at a constant level. The amount of sulfates bound by the mortar mixtures and any change in the mass and volume of the samples were measured at regular intervals. In addition, dynamic modulus of elasticity, compressive and bending strength measurements were also regularly performed.

Based on the analysis of the results obtained during the first four years of the test program, the author proposed the following equation to predict the degree of sulfate-induced corrosion (DC):

$$DC = [0.11S^{0.45}] [0.143t^{0.33}] [0.204e^{0.145C_3A}] \qquad (7.1)$$

where $S$ stands for the $SO_4$ concentration of the test solution (expressed in mg/l), $t$ is the immersion period (expressed in days) and $C_3A$ is the percentage in tricalcium aluminate of the Portland cement (calculated according to Bogue's equations).

It should be emphasized that the degree of corrosion predicted by equation (7.1) mainly describes the amount of sulfates bound by the solid over time. Bound sulfate results were found by the author to correlate well with volume change data.

The author also proposes to multiply equation (7.1) by a correcting term ($\eta_a$) to account for the presence of supplementary cementing materials (such as slag and the *volcanic tuff*):

$$\eta_a = e^{-0.0164} \qquad (7.2)$$

where $A$ represents the level of replacement of the Portland cement by the supplementary cementing material (expressed as a percentage of the total mass of binder). This correcting term was calculated on the basis of a series of experimental results summarized in Figure 7.1.

As can be seen, the degree of corrosion predicted by Jambor's model (equations (7.1) and (7.2)) is directly affected by the sulfate concentration of the test solution and the $C_3A$ content of the cement used in the preparation of the mixture. This is in good agreement with most empirical equations found in the literature. In that respect, the model is useful to investigate the influence of various parameters (such as cement composition) on the behavior of laboratory samples. It is, however, difficult to predict the service-life of concrete structures solely on the basis of Jambor's model. The author does not provide any information on the critical degree of corrosion beyond which the service-life of a structure is compromised.

According to Jambor's model, the DC does not evolve linearly with time. As will be seen in the following paragraphs, this is in contradiction with other empirical models recently proposed by various authors. The non-linear nature of Jambor's model can probably be explained by the fact that the validity of equations (7.1) and (7.2) is limited to samples fully immersed in the test solutions. Under these conditions, sulfate ions mainly penetrate by

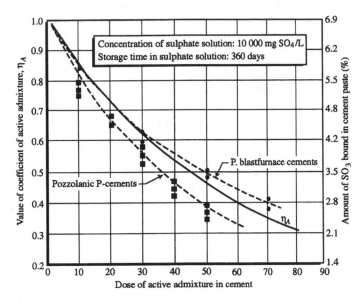

*Figure 7.1* Relationship between the dose of active mineral admixture and the degree of corrosion of samples exposed for 360 days to a sulfate solution (10,000 mg of $SO_4$ per liter).
*Source*: Jambor (1998)

diffusive process that can be approximated by a non-linear relationship. Furthermore, Jambor's model does not take explicitly into account the influence of the microstructural damage induced to the material on the kinetics of sulfate penetration. This effect is only implicitly considered in the second term of equation (7.1).

As emphasized by the author himself, equations (7.1) and (7.2) are only valid for mortar mixtures prepared at a water–binder ratio of 0.6 and fully immersed in sodium sulfate solutions maintained at a constant concentration and constant temperature (in this case 20 °C). These equations do not account for any variations of the test conditions, neither can they be used to assess the influence of various parameters (such as water–binder ratio or time of curing) on the sulfate resistance of the mixture. Finally, these equations cannot obviously serve to predict the durability of hydrated cement systems exposed to calcium sulfate or magnesium sulfate solutions.

Numerous empirical models, similar to that of Jambor, have been developed over the years. Since most of them have been extensively reviewed by Clifton (1991), only a brief description of these various models will be given in the following paragraphs.

Probably the best known of these empirical models is the equation proposed by Atkinson and Hearne (1984). This model is derived from an analysis of the laboratory data obtained by Harrison and Teychenne (1981) who tested various concrete samples fully immersed in a 0.19 M sulfate solution (a mixture of sodium sulfate and magnesium sulfate) over a five-year period. Based on these data, Atkinson and Hearne (1984) developed the following equation to predict the location $(X_s)$ of the visible degradation zone:

$$X_s \text{ (cm)} = 0.55C_3A \cdot ([Mg] + [SO_4]) \cdot t(y) \qquad (7.3)$$

where $C_3A$ stands for the tricalcium aluminate content of the cement (expressed as a percentage of the mass of cement), [Mg] and [$SO_4$] are the molar concentrations in magnesium and sulfates, respectively, in the test solution, and $t(y)$ is the immersion period in years.

As can be seen, contrary to the model of Jambor (1998), the equation proposed by Atkinson and Hearne (1984) predicts that the sulfate-induced degradation will evolve as a linear function of time. This contradiction between the two models is particularly important since the application of both equations is limited to samples fully immersed in solution.

It should also be emphasized that neither equation (7.3) nor Jambor's model takes into account the influence of water–cement (or water–binder) of concrete on the kinetics of degradation. This limitation of equation (7.3) was later acknowledged by Atkinson and Hearne (1990).

The equation was found to give satisfactory correlation with the results of field tests, in which the depths of penetration were in the range of 0.8–2 cm after five years. The equation was also used by the authors to calculate the service life of concrete samples exposed to ground water of a known sulfate concentration. Concrete made with ordinary Portland cements containing 5–12% $C_3A$, gave estimated lifetimes of 180–800 years, with a probable lifetime of 400 years. When sulfate resisting Portland cement with 1.2% $C_3A$ was used, the minimum and probable lifetimes were estimated to be 700 years and 2,500 years, respectively. These times were estimated based on the loss of one-half of the load-bearing capacity of a 1-m thick concrete section, i.e., $X_s$ of 50 cm.

Atkinson *et al.* (1986) also attempted to validate equation (7.3) by determining the extent of deterioration of concretes buried in clay for about forty years. An alteration zone of about 1 cm was observed in the samples. However, the authors mentioned that sulfate attack was probably not the only cause of degradation. Based on the tricalcium aluminate contents of the cements, equation (7.3) predicts that the thickness of the deteriorated region should be between 1 and 9 cm. Therefore, the authors concluded that the equation was slightly overestimating the rate of sulfate attack.

A modification of the Atkinson and Hearne (1984) model was later proposed by Shuman *et al.* (1989). According to this model, the thickness of the degraded zone can be estimated using the following equation:

$$X_s = 1.86 \times 10^6 \, C_3A(\%) \cdot ([Mg] + [SO_4])D_i \cdot t \qquad (7.4)$$

where $D_i$ is the apparent diffusion coefficient of sulfate ions in the material. As can be seen, the main difference between expressions (7.3) and (7.4) is that a correction is made to the latter to account for the diffusion coefficient of the mixture.

As for the two previous models, the expression proposed by Shuman *et al.* (1989) does not explicitly consider any influence of the water–cement ratio of the material on the rate of degradation. However, the effect of the mixture characteristics is indirectly taken into consideration by the diffusion coefficient ($D_t$). Apparently, Shuman *et al.* (1989) have not attempted to perform any experimental validation of their model.

Rasmuson and Zhu (1987) developed another model in which the rate of degradation is directly affected by the diffusion of sulfate ions into the material. In this approach, sulfate ions move through degraded concrete to the interface of unreacted concrete, and then react with the hydration products of tricalcium aluminate to form expansive products such as ettringite. Mass transport equations are used, assuming a quasi-steady state, to predict the movement of sulfates in the concrete. The flux of sulfate ions, $N$, is given by:

$$N = -D_i \left( \frac{C_0}{x} \right) \tag{7.5}$$

where $C_0$ is the concentration (in mol/l) of sulfate in the bulk solution, $D_i$ is the intrinsic diffusion coefficient of sulfate ions into the material (in $m^2/s$), and $x$ is the depth of degradation (in meters).

The rate of deterioration is essentially controlled by the rate of mass transport divided by the $C_3A$ content of the material:

$$\frac{dx}{dt} = -\frac{N}{C_a} = D_i \frac{C_0}{C_a} x \tag{7.6}$$

In agreement with the previous empirical expressions, the model predicts that the rate of sulfate attack decreases with increasing amounts of $C_3A$.

More recently, a series of two empirical equations were proposed by Kurtis *et al.* (2000) to predict the behavior of concrete mixtures partially submerged in a 2% (0.15 M) sodium sulfate solution.[1] The two expressions were derived from a statistical analysis of a total of 8,000 expansion measurements taken over a forty-year period by the US Bureau of Reclamation on 114 cylindrical specimens ($76 \times 152$ mm). The two equations are based on results collected from fifty-one different mixtures with w/c ranging from 0.37 to 0.71 and including cements with $C_3A$ contents ranging from 0 to 17%.

The statistical analysis of the data clearly revealed a disparity in performance between the cylinders produced with low (i.e. <8%) and high (>10%) $C_3A$ contents. This phenomenon prompted the authors to propose an empirical equation for each category of mixtures. Hence, the authors developed the following expression to predict the expansion (Exp, expressed in percent) of concrete mixtures made of cement with low (i.e. <8%) $C_3A$ content:

$$Exp = 0.0246 + [0.0180(t)(w/c)] + [0.00016(t)(C_3A)] \tag{7.7}$$

where time ($t$) is expressed in years. In the equation, w/c stands for the water–cement ratio of the mixture and $C_3A$ corresponds to the tricalcium aluminate content of the cement (in per cent). According to the authors, this equation should be valid for w/c in the 0.37–0.71 range and for severe sulfate exposure up to forty years.

The following equation was proposed for concrete mixtures prepared with cement with a high (>10%) $C_3A$ content:

$$\ln(Exp) = -3.753 + [0.930(t)] + [0.0998 \ln((t)(C_3A))] \tag{7.8}$$

According to the authors, the latter equation should be considered for w/c in the range 0.45–0.51 for severe sulfate exposure up to forty years. Typical examples of the application of these equations are given in Figures 7.2 and 7.3. The two equations proposed by Kurtis *et al.* (2000) appear to form one of the most complete empirical models developed over the years. As can be seen, their equations consider the influence of two critical mixture characteristics: $C_3A$ content of the cement and water–cement ratio (at least equation (7.7) which was developed for a wide range of mixtures). The two equations

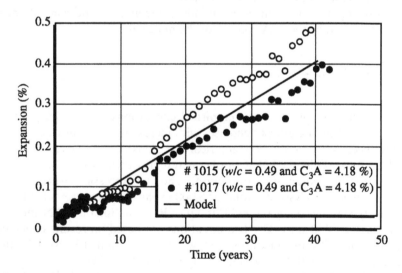

*Figure 7.2* Model prediction (equation 7.7) for concrete mixtures with w/c 0.49 and $C_3A$ content of 4.18% and expansion data for two specimens with same characteristics.
*Source*: Kurtis *et al.* (2000)

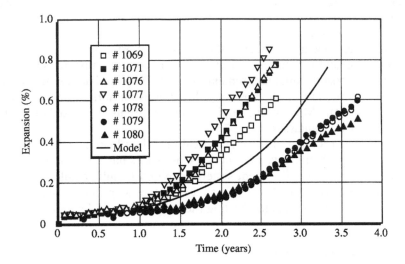

*Figure 7.3* Model prediction (equation 7.8) for concrete mixtures with $C_3A$ content of 17% and expansion data for seven specimens with w/c between 0.46 and 0.47.
*Source*: Kurtis *et al.* (2000)

were also derived for concrete samples partially immersed in solution which corresponds to the conditions most commonly found in service. Unfortunately, the models do not take into account the effect of the sulfate concentration of the surrounding solution nor the influence of different types of sulfate solutions (such as magnesium sulfate and calcium sulfate).

It should be emphasized that, contrary to the previous approaches, the two equations proposed by Kurtis *et al.* (2000) can be used to calculate the expansion of concrete cylinders. One cannot rely on them to predict, as in the previous models, the rate of penetration of the sulfate degradation layer. It is also interesting to note that the kinetics of expansion predicted by the two equations tend to differ according to the $C_3A$ content of the cement used in the preparation of the concrete mixture.

Equation (7.7) (valid for a wide range of concrete mixtures) also provides some interesting information on the relative importance of $C_3A$ content and water–cement ratio on the durability of concrete exposed to sulfate-rich environment. According to this expression, the latter parameter has clearly a strong influence on the behavior of concrete. For instance, equation (7.7) indicates that an increase of the water–cement ratio from 0.45 to 0.70 should increase by approximately 40% the ten-year expansion of a concrete mixture prepared with a cement containing 4% of $C_3A$. Similarly, an increase of the $C_3A$ content from 4 to 8% should increase by only 10% the ten-year expansion of a 0.45 water–cement ratio concrete mixture.

The previous example clearly illustrates the main advantage of most empirical models. The influence of a single parameter on the behavior of the material can simply be evaluated on the basis of a relatively straightforward calculation. Furthermore, calculations can usually be performed using a limited number of input data.

Despite these clear advantages, the ability of most empirical models to accurately predict the behavior of a wide range of concrete mixtures subjected to different exposure conditions remains limited. These limitations are usually not linked to the approach chosen by the various authors to analyze their experimental data. Most recent empirical models are usually based on sophisticated statistical analyses.

The intrinsic problem of these empirical models is linked to the complex nature of the problem. Given the number of factors having an influence on the behavior of hydrated cement systems exposed to sulfate solutions, it is practically impossible to carry out an experimental program that would encompass all the parameters affecting the mechanisms of degradation.

### 7.2.2    Mechanistic models

More recently, researchers have tried to develop a new generation of more sophisticated models to predict the service life of concrete exposed to sulfate environments. These mechanistic (or phenomenological) models can be distinguished from the purely empirical equations by the fact that they are generally based on a better understanding of the mechanisms involved in the degradation process. However, since many of these mechanistic models rely, to a great extent, on empirically based coefficients, the line separating these two categories is often thin.

Being aware of the intrinsic limitations of their empirical model, Atkinson and Hearne (1990) were probably the first authors to develop a mechanistic model for predicting the effect of sulfate attack on service life of concrete. The model is based on following assumptions:

1    Sulfate ions from the environment penetrate the concrete by diffusion;
2    Sulfate ions react expansively with aluminates in the concrete; and
3    Cracking and delamination of concrete surfaces result from these expansive reactions.

The model predicts that rate of surface attack will be largely controlled by the concentration of sulfate ions and aluminates, diffusion and reaction rates, and the fracture energy of concrete. One important feature of this model is that the authors did not assume the existence of a local chemical equilibrium between the diffusing sulfate ions and the various solid phases within the material. The kinetics of reaction is rather described by an empirical equation derived from immersion experiments of a few grams of hydrated cement paste in sulfate solutions. Typical curves obtained from two of these immersion tests are given in Figure 7.4.

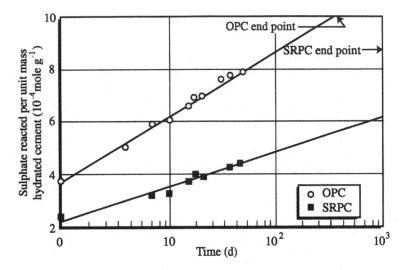

*Figure 7.4* The reaction kinetics of hydrated suspensions of OPC and SRPC with sulfate from a saturated solution of gypsum in lime water (sufate concentration 12.2 mM).

*Source*: Atkinson and Hearne (1990)

Another important feature of the mechanistic model is that the maximum amount of sulfate that can be bound by the solid is also estimated on the basis of immersion test results. This approach allows taking into account the influence of all the various sources of aluminate found in a hydrated cement system. In the previous empirical model (Atkinson and Hearne 1984), the amount of bound sulfates was exclusively controlled by the $C_3A$ content of cement.

The authors also developed additional relationships for the thickness of concrete, which spalls, the time for a layer to spall, and the degradation rate. The degradation rate ($R$) is linear in time (m/s) and is given by:

$$R = \frac{X_{\text{spall}}}{T_{\text{spall}}} = \frac{(EB^2 C_s C_0 D_i)}{\alpha \tau (1 - \nu)} \tag{7.9}$$

where   $X_{\text{spall}}$ is the thickness of a spalled layer,
$T_{\text{spall}}$ is the time for a layer to spall,
$E$ is Young's modulus,
$B$ is the linear strain caused by one mole of sulfate, reacted in $1\,\text{m}^3$ of concrete,
$C_s$ is the sulfate concentration in bulk solution,
$C_0$ is the concentration of reacted sulfate as ettringite,
$D_i$ is the intrinsic diffusion coefficient of sulfate ions,
$\alpha$ is a roughness factor for fracture path (assumed to be equal to 1),

$\tau$ is the fracture surface energy of concrete, and
$\nu$ is Possion's ratio.

Some of the data needed to solve the model need to be obtained from laboratory experiments. Other parametric data to solve the model are not available for specific concretes and typical values must be used.

As can be seen, the model by Atkinson and Hearne (1990) predicts that the diffusion coefficient and the sulfate concentration of the ground water are the most significant factors controlling the resistance of concrete to sulfate attack. As emphasized by Clifton (1991) in his comprehensive review of service-life prediction models, the mechanistic approach proposed by Atkinson and Hearne (1990) gives the same time order as their previous empirical model. However, in constrast to the previous empirical equation, the mechanistic model can be applied to a wider range of concrete mixtures.

No attempt to correlate the model to experimental data was reported by Atkinson and Hearne (1990). However, the authors mention that, since the model neglects visco-elastic effects (that should contribute to minimize cracking in the reaction zone), equation (7.9) probably overestimates the rate of degradation of concrete.

Another mechanistic model was later developed by Clifton and Pommersheim (1994) to predict the volumetric expansion of cementitious materials as a function of the specific expansive chemical reaction, degree of hydration, the composition of the concrete, and the densities of the individual phases. The model is mainly based on the concept of excluded volume, whereby, the amount of expansion is presumed to be proportional to the difference between the net solid volume produced and the original capillary porosity.

More specifically, the model has been developed to predict the volumetric expansion upon sulfate attack. The model is based on the potential for expansion provided by both $C_3A$ content of the cement and the sulfate ion concentration of penetrating aqueous solutions. It also considers the amount of cement in concrete and the characteristics of pores in which expansive products of the reactions can grow.

The mathematical model, which predicts the fractional expansion, $X$ of cementitious materials exposed to sulfate solution is given by the following equations:

$$X = h(X_p - \phi_c), \quad X_p > \phi_c \tag{7.10}$$

$$X = 0, \quad X_p \le \phi_c \tag{7.11}$$

where $\phi_c$ is the capillary porosity of the concrete. The constant $h$ is introduced to account for the degree to which the potential expansive volume, as measured by $(X_p - \phi_c)$, is translated into actual expansion. If $h = 1$ all of the expansive products would cause expansion, while for $h < 1$ only some of the potentially expansive products would create expansion. In the model, the value of *h* is assumed to be equal to 0.05.

Using this model, Clifton and Pommersheim (1994) could establish that:

1   Ettringite formation from monosulfate is not likely to cause expansion. However, local expansion could occur if ettringite occupied the same pore space left by reacting monosulfate;
2   If sufficient unhydrated tricalcium aluminate is available in mature cement paste and it reacts with gypsum to form ettringite, the reaction will likely lead to expansion at low w/c ratios;
3   The conversion of calcium hydroxide to gypsum is not likely to produce any significant expansion.

As can be seen, mechanistic models provide useful tools to investigate the parameters that control the resistance of concrete to sulfate attack. Their main advantage lies on the fact that they are based on a much better understanding of the fundamental mechanisms that control the degradation of concrete exposed to sulfates. However, as for empirical models, their ability to reliably predict the service-life of concrete structures remains somewhat limited.

### 7.2.3   Numerical models

In the past decade, computers have been increasingly involved in microstructure-based modeling. As previously mentioned, the development of new numerical methods and the constant improvement of the field of computer engineering have encouraged engineers and scientists to elaborate more sophisticated models. Over the past decades, many of these models have been devoted to the prediction of the service life of concrete structures exposed to chemically aggressive environments.

Snyder and Clifton (1995) were among the first authors to develop a numerical model specifically dedicated to the prediction of the behavior of concrete subjected to external sulfate attack. This model, called 4SIGHT, can be used to predict the chemical degradation of reinforced concrete structures subjected to various aggressive chemical species (e.g. sulfates, chlorides, etc.). In this model, degradation mechanisms are typically controlled by the concentration of ions in the pore solution. Ionic species are propagated through concrete using the following advection–diffusion equation:

$$\mathbf{j} = -D\partial c/\partial x + c\mathbf{u} \tag{7.12}$$

where $\mathbf{j}$ is the ionic flux, $c$ the concentration, $D$ the effective diffusion coefficient and $\mathbf{u}$ the average pore fluid velocity.

After every step of ionic transport, each computational concrete element is put in chemical equilibrium using solubility products and charge balance equations. Although the model does not explicitly take into consideration the electrostatic coupling between the various ionic fluxes, ionic concentrations are corrected in order to maintain the local electroneutrality of the solution. This step is achieved through dissolution–precipitation of available salts. Chemical activity effects are not taken into account by the model. However, an interesting feature of the approach is that the influence of chemical reactions on the porosity of the material (and its transport properties) is taken into consideration.

The model relies of approach proposed by Atkinson and Hearne (1984) to predict the degradation thickness of concrete upon a sulfate attack. Hence, the rate of degradation is calculated using equation (7.9). According to this approach, the kinetics of concrete degradation is not related to the local chemistry of the pore solution within the material but rather to the sulfate concentration at the vicinity of the surface of concrete. Unfortunately, the model was not developed to treat any fluctuations of the boundary conditions (at the concrete surface) over time (e.g. wetting and drying or freezing and thawing cycles can not be considered). This can be explained by the fact that the model was originally intended to predict the service life of low level waste disposal facilities (i.e. where the boundary conditions are virtually always constants).

The main advantage of this model is that it can be used to predict both the distribution of sulfate-bearing phases and the residual mechanical properties of concrete subject to sulfate attack. In that respect, 4SIGHT is probably one of the most complete model so far published. Unfortunately, the model has not been the subject of a systematic validation.

Another numerical model was later developed by Marchand *et al.* (1999). This latter model has been developed to predict the transport of ions in unsaturated porous media in isothermal conditions. The model also accounts for the effect of dissolution–precipitation reactions on the transport mechanisms.

The description of the various transport mechanisms relies on the homogenization technique. This approach first requires writing all the basic equations at the microscopic level. These equations are then averaged over a Representative Elementary Volume (REV) in order to describe the transport mechanisms at the macroscopic scale (Bear and Bachmat 1991; Samson *et al.* 1999a).

In this model, ions are considered to be either free to move in the liquid phase or bound to the solid phase. The transport of ions in the liquid phase at the microscopic level is described by the extended Nernst–Planck equation to which is added an advection term. After integrating this equation over the REV, the transport equation becomes:

$$(1 - \phi)\frac{\partial c_{is}}{\partial t} + \frac{\partial(\theta c_i)}{\partial t} - \frac{\partial}{\partial x}\left(\theta D_i\frac{\partial c_i}{\partial x} + \theta\frac{D_i z_i F}{RT}c_i\frac{\partial \psi}{\partial x} - c_i V_x\right) = 0 \qquad (7.13)$$

where   $c_i$ is the concentration of the species i (mol/l),

$c_{is}$ is the concentration in ions bound to the solid (mol/m$^3$)

$D_i$ is the diffusion coefficient of the species i (m$^2$/s),

$D_w$ is the diffusion coefficient of water (m$^2$/s),

$F$ is the Faraday constant (9.64846E04 Coulomb/mol),

$R$ is the ideal gas constant,

$T$ is the temperature (°K),

$z_i$ is the ion valency,

$\theta$ is the liquid water content (m$^3$/m$^3$ of concrete), and

$\psi$ the diffusion potential set up by the drifting ions (in Volt).

Equation (7.13) has to be written for each ionic species present in the system. To calculate the chemical activity coefficients, several approaches are available. However, models such as those proposed by Debye–Hückel or Davies are unable to reliably describe the thermodynamic behavior of highly concentrated electrolytes such as the hydrated cement paste pore solution. A modification of the Davies equation described by Samson *et al.* (1999b) was found to yield good results.

The Poisson equation is added to the model to evaluate the electrical potential ($\psi$). It relates the electrical potential to the concentration of each ionic species. The equation is given here in its averaged form:

$$\frac{d}{dx}\left(\theta\tau\,\frac{d\psi}{dx}\right) + \theta\,\frac{F}{\varepsilon} + \sum_{i=1}^{N} z_i c_i = 0 \qquad (7.14)$$

where $N$ is the total number of ionic species, $\varepsilon$ is the dielectric permittivity of the medium, in this case water, and $\tau$ is the tortuosity of the porous network.

The velocity of the fluid, appearing in equation (7.13) as $V_x$, can be described by a diffusion equation when its origin is in capillary forces present during drying–wetting cycles:

$$V_x = -D_w\,\frac{\partial\theta}{\partial x} \qquad (7.15)$$

where $D_w$ is the non-linear water diffusion coefficient. This parameter varies according to the water content of the material (Pel 1995).

To complete the model, the mass conservation on the liquid phase must be taken into account:

$$\frac{\partial\theta}{\partial t} - \frac{\partial}{\partial x}\left(D_w\,\frac{\partial\theta}{\partial x}\right) = 0 \qquad (7.16)$$

As can be seen, moisture transport is described in terms of variation of the water content (liquid) of the material. It should be emphasized that the choice of using the material water content as the state variable for the description of this problem has an important implication on the treatment of the boundary conditions. Since the latter are usually expressed in terms of relative humidity, a conversion has to be made. This can be done using an adsorption–desorption isotherm (Pel 1995).

The first term on the left-hand side of equation (7.13) (in which $c_{is}$ appears), accounts for the ionic exchange between the solution and the solid. It can be used to model the influence of precipitation–dissolution reactions on the transport process. The chemical equilibrium of the various solid phases present in the material is verified at each node by considering the concentrations of all ionic species at this location. If the equilibrium condition is not respected, the concentrations and the solid phase content are corrected accordingly using a chemical equilibrium code. More information on this procedure can be found in Marchand (2001).

The influence of on-going chemical reactions on the material transport properties is accounted for. The effects of the chemically induced alterations are described in terms of porosity variations.

The transport of ions and water in unsaturated cement systems can be fully described on the basis of equations (7.13)–(7.16). Previous experience (Marchand 2001) has shown that most practical problems can be reliably described by seven different ionic species ($OH^-$, $Ca^{2+}$, $Na^+$, $K^+$, $Cl^-$, $SO_4^{2-}$, $Al(OH)_4^-$) and six solid phases (CH, C-S-H, ettringite, gypsum, chloro aluminates and hydrogarnet).

The input data required to run the model can be easily obtained. The initial composition of the material (i.e. its initial content in CH, ettringite, etc.) can be easily calculated by considering the chemical (and mineralogical) make-up of the binder, the characteristics of the mixture and the degree of hydration of the system.

The model also requires determining the initial composition of the pore solution and the porosity of the material. Samples of the pore solution of most hydrated cement systems can be obtained by extraction. The total porosity of the material can easily be determined in the laboratory following standardized procedures (such as ASTM C642).

Some information on the transport properties of the material is also required to run the model. The ionic diffusion properties of the solid can be determined by a migration test. The water diffusion coefficient of the material can be assessed by nuclear magnetic resonance imaging.

The model can be used to follow any changes in the concrete pore solution chemistry. It also provides a precise description of the material solid phase distribution. This model has been the subject of a very systematic experimental validation. For instance, it has been successfully applied to calcium leaching and sulfate degradation experiments (Maltais *et al.* 2001; Marchand 2001; Marchand *et al.* 1999; Marchand *et al.* 2001). Typical applications of the model are given in Figures 7.5 and 7.6.

The model has also been used to predict the behavior of field concrete exposed to sulfate-bearing soils. Numerical simulations clearly emphasized

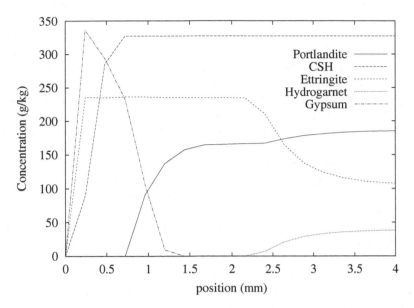

*Figure 7.5* Distribution in solid phases for a 0.6 w/c ratio mixture made of an ASTM
Type 1 cement and immersed in a solution of $Na_2SO_4$ at 50 mmol/l.
*Source*: Marchand (2001)

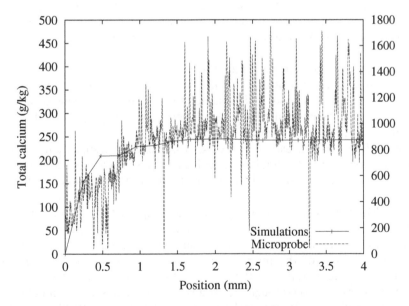

*Figure 7.6* Total calcium profile for a 0.6 w/c ratio mixture made of an ASTM Type 1
cement and immersed in a solution of $Na_2SO_4$ at 50 mmol/l.
*Source*: Marchand (2001)

the primary importance of water–cement ratio on the performance of concrete.

A similar numerical model was recently proposed by Schmidt-Döhl and Rostásy (1999). This latter approach essentially consists in predicting the transport of ions and water in unsaturated concrete in isothermal conditions. The algorithm used by the authors (see Figure 7.7) is inspired from the work of van Zeggeren and Storey (1970) and differs significantly from that developed by Marchand et al. (1999). In addition, the effects of chemical reactions on the transport properties of the material are not taken into consideration. Furthermore, the influence of the diffusion potential on the transport of ions is not calculated using Poisson's equation but is rather indirectly taken into account through a correction term. The model was found to reliably predict the penetration of sulfate ions into mortar samples. Typical results are given in Figure 7.8.

As can be seen, engineers can now rely on numerical models to obtain reliable descriptions of the microstructural alterations of hydrated cement systems subjected to sulfate attack. Unfortunately, these numerical models do not provide (at least not directly) any information on the consequences of the degradation on the mechanical properties of the material.

Over the past decades, numerous authors have also attempted to develop numerical models to predict the effect of sulfate attack on the mechanical properties of concrete. One typical example is the model proposed by Ouyang (1989) to investigate the behavior of mortar mixtures fully immersed in sodium sulfate solutions. Based on the isotropic damage theory and the progressive fracturing concept, the model takes into consideration the initial porosity of the material and the subsequent influence of sulfate attack on the pore structure.

In this model, the damage variable is defined as the ratio of void area to the cross-sectional area of the sample. The "initial" damage of the system (i.e. that exists prior to loading) is assumed to be induced by drying shrinkage and various environmental effects, and is believed to be proportional to the gel space ratio defined by Powers (1958).

The sulfate attack measured by the expansion of a given specimen is incorporated into the model as additional nucleated voids. The stress–strain curves are then predicted by using a damage growth law and a uniaxial constitution equation.

The algorithm used by author is presented in Figure 7.9 and typical results are given in Figure 7.10. It should be emphasized that the purpose of the model is not to predict the evolution of sulfate attack but rather to predict the consequences of the microstructural alterations on the mechanical properties of the material. In that respect, the kinetics of the problem are taken into consideration through the expansion curve that is used as input data.

More recently, a similar approach was proposed by Ju et al. (1999). The latter model is based on a linear damage accumulation law for sulfate attack induced damage. According to this approach, the rate of damage is assumed

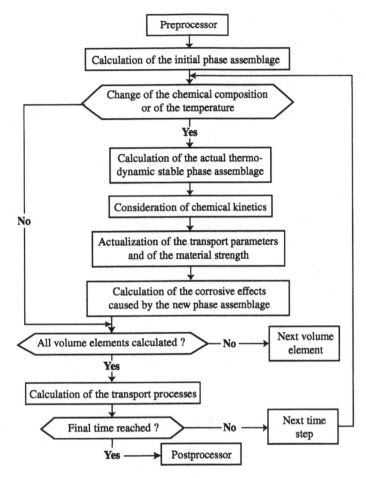

*Figure 7.7* Structure of the numerical algorithm.
*Source*: Schmidt-Döhl and Rostasy (2000)

to be constant over the service life of the structure, and failure is defined by considering the evolution of the tensile strength of the material. This approach has served to investigate the mechanical properties of a certain number of field concrete samples that had been exposed to sulfate-bearing soils.

## 7.3 CONCLUDING REMARKS

As can be seen, numerous different models have been developed to predict the resistance of concrete to sulfate attack. Although many empirical and mechanistic models tend to yield fairly reliable results, only numerical models were found to have to ability to capture the complex nature of the degradation mechanisms.

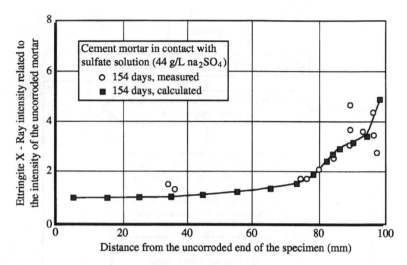

*Figure 7.8* Ettringite profile of mortar samples corroded by $Na_2SO_4$ solution (44 g/l).
*Source*: Schmidt-Döhl and Rostasy (2000)

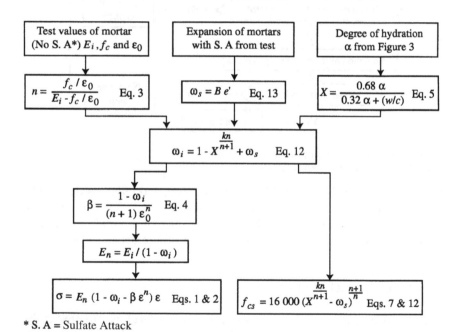

$$n = \frac{f_c / \varepsilon_0}{E_i - f_c / \varepsilon_0} \quad \text{Eq. 3}$$

$$\omega_s = B\,e' \quad \text{Eq. 13}$$

$$X = \frac{0.68\,\alpha}{0.32\,\alpha + (w/c)} \quad \text{Eq. 5}$$

$$\omega_i = 1 - X^{\frac{kn}{n+1}} + \omega_s \quad \text{Eq. 12}$$

$$\beta = \frac{1 - \omega_i}{(n+1)\,\varepsilon_0^n} \quad \text{Eq. 4}$$

$$E_n = E_i / (1 - \omega_i)$$

$$\sigma = E_n\,(1 - \omega_i - \beta\,\varepsilon^n)\,\varepsilon \quad \text{Eqs. 1 \& 2}$$

$$f_{cs} = 16\,000\,(X^{\frac{kn}{n+1}} - \omega_s)^{\frac{n+1}{n}} \quad \text{Eqs. 7 \& 12}$$

* S. A = Sulfate Attack

*Figure 7.9* Structure of the numerical algorithm.
*Source*: Ouyang (1989)

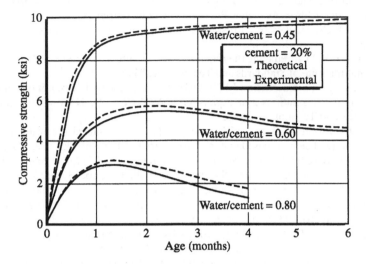

*Figure 7.10* Comparison of experimental and theoretical compressive strength for mortars with different w/c ratio under external sulfate attack.

*Source*: Ouyang (1990)

It should, however, be emphasized that most of the actual numerical models are limited to the prediction of one aspect of the degradation process. More work is required to develop a model that will be able to reliably predict the microstructural alterations induced to the material by sulfate attack and their consequences on the mechanical properties of the structure.

## NOTE

1  This concentration corresponds to a severe sulfate exposure.

## REFERENCES

Atkinson, A., Goult, D.J. and Hearne, J.A. (1986) "An assessment of the long-term durability of concrete in radioactive waste repositories", *Materials Research Society Symposium Proceedings* **50**: 239–246.

Atkinson, A. and Hearne, J.A. (1984) "An assessment of the long-term durability of concrete in radioactive waste repositories", *AERE-R11465*, Harwell, UK.

Atkinson, A. and Hearne, J.A. (1990) "Mechanistic model for the durability of concrete barriers exposed to sulphate-bearing groundwaters", *Materials Research Society Symposium Proceedings* **176**: 149–156.

Bear, J. and Bachmat, Y. (1991) *Introduction to modeling of transport phenomena in porous media*, Kluwer Academic Publishers, The Netherlands.

Biczok, I. (1967) "Concrete corrosion–concrete protection", Chemical Publishing Inc., New York, pp. 138–143.

Clifton, J.R. (1991) "Predicting the remaining service-life of concrete", *NISTIR 4712*, National Institute of Standards and Technology, 73 pp.

Clifton, J.R. and Pommersheim, J.M. (1994) "Sulfate attack of cementitious materials: Volumetric relations and expansions", *NISTIR 5390*, National Institute of Standards and Technology, 20 pp.

Clifton, J.R., Frohnsdorff, G. and Ferraris, C. (1999) "Standard for evaluating the susceptibility of cement-based materials to external sulfate attack", in J. Marchand and J. Skalny (eds) *Materials Science of Concrete Special Volume: Sulfate Attack Mechanisms*, American Ceramic Society, Westerville OH, pp. 337–355.

Figg, J. (1999) "Field studies of sulfate attack on concrete", in J. Marchand and J. Skalny (eds) *Materials Science of Concrete Special Volume: Sulfate Attack Mechanisms*, The American Ceramic Society, Westerville, OH, pp. 315–324.

Garboczi, E.J. (1990) "Permeability, diffusivity and microstructural parameters: A critical review", *Cement and Concrete Research* **20**: 591–601.

Garboczi, E.J. (2000) Presentation made at the J.F. Young Symposium, May.

Harboe, E.M. (1982) "Longtime studies and field experiences with sulfate attack", in *Sulfate Resistance of Concrete* (George Verbeck Symposium), ACI SP-77, pp. 1–20.

Harrison, W.H. and Teychenne, D.C. (1981) "Sulphate resistance of buried concrete", in *Second Interim Report on Long Term Investigation at Northwick Park*, Building Research Establishment, Garston, UK.

Jambor, J. (1998) "Sulfate Corrosion of Concrete", an unpublished manuscript summarizing Dr Jambor's work on the *Sulfate durability of concrete*. (The author passed away in May 1998.)

Ju, J.W., Weng, L.S., Mindess, S. and Boyd, A.J. (1999) "Damage assessment and service-life prediction of concrete subject to sulfate attack", in J. Marchand and J. Skalny (eds) *Materials Science of Concrete Special Volume: Sulfate Attack Mechanisms*, The American Ceramic Society, Westerville, OH, pp. 265–282.

Kalousek, G.L., Porter, L.C. and Benton, E.J. (1972) "Concrete for long-time service in sulfate environment", *Cement and Concrete Research* **2**: 79–89.

Kurtis, K.E., Monteiro, P.J.M. and Madanat, S. (2000) "Empirical models to predict concrete expansion caused by sulfate attack", *J. ACI Materials*, March–April 2000, V97: 156–161. Errata published November–December 2000, V97: 713.

Maltais, Y., Marchand, J. and Samson, E. (2001) "Predicting the behavior of hydrated cement systems subjected to concentrated sulfate solutions" (in preparation).

Marchand, J. (2001) "Modeling the behavior of unsaturated cement systems exposed to aggressive chemical environments", *Materials and Structures* **34**: 195–200.

Marchand, J., Gérard, B. and Delagrave, A. (1998) "Ion transport mechanisms in cement-based materials", in J. Skalny and S. Mindess (eds) *Materials Science of Concrete*, vol. V. The American Ceramic Society, pp. 307–400.

Marchand, J., Samson, E. and Maltais, Y. (1999) "Modeling microstructural alterations of concrete subjected to sulfate attack", in J. Marchand and J. Skalny (eds) *Materials Science of Concrete Special Volume: Sulfate Attack Mechanisms*, The American Ceramic Society, Westerville, OH, pp. 211–257.

Marchand, J., Bentz, D.P., Samson, E. and Maltais, Y. (2001) "Influence of calcium hydroxide dissolution on the transport properties of hydrated cement systems", in J. Skalny, J. Gebauer and I. Odler (eds) *Materials Science of Concrete Special Volume: Calcium Reactions in Concrete,* The American Ceramic Society, Westerville, OH (in press).

Nilsson, L.O., Poulsen, E., Sandberg, P., Sorensen, P. and Klinghoffer, O. (1996) "HETEK – Chloride penetration into concrete", *Report No. 53*, The Road Directorate, Copenhagen, Denmark.

Ouyang, C. (1989) "A damage model for sulfate attack of cement mortars", *Cement, Concrete, and Aggregates*, CCAGDP, **11**(2): 92–99.

Pel, L. (1995) "Moisture transport in porous building materials", Ph. D. Thesis, Eindhoven University of Technology, The Netherlands, 125 pp.

Powers, T.C. (1958) "Structure and physical properties of hardened Portland cement paste", *Journal of the American Ceramic Society* **41**: 1–6.

Rasmuson, A. and Zhu, M. (1987) "Calculations of the degradation of concrete in a final repository for nuclear water", in *Proceeding of NEA Workshop on Near-Field Assessment of Repositories for Low and Medium Level Radioactive Wastes*, Baden, Switzerland.

Reinhardt, H.W. (1996) "Transport of chemicals through concrete", in J. Skalny (ed.) *Materials Science of Concrete*, vol. III, The American Ceramic Society, Westerville, Ohio, USA, pp. 209–242.

Samson, E., Marchand, J. and Beaudoin, J.J. (1999a) "Describing ion diffusion mechanisms in cement-based materials using the homogenization technique", *Cement and Concrete Research* **29**: 1341–1345.

Samson, E., Lemaire, G., Marchand, J. and Beaudoin, J.J. (1999b) "Modeling chemical activity effects in strong ionic solutions", *Computational Materials Science* **15**: 285–294.

Schmidt-Döhl, F. and Rostásy, F.S. (1999) "A model for the calculation of combined chemical reactions and transport processes and its application to the corrosion of mineral-building materials – Part I: Simulation model", *Cement and Concrete Research* **29**: 1039–1045.

Schmidt-Döhl, F. and Rostásy, F.S. (1999) "A model for the calculation of combined chemical reactions and transport processes and its application to the corrosion of mineral-building materials – Part II: Experimental verification", *Cement and Concrete Research* **29**: 1047–1053.

Shuman, R., Rogers, V.V. and Shaw, R.A. (1989) "The Barrier Code for predicting long-term concrete performance", *Waste Management* **89**, University of Arizona.

Smith, F.L. (1958) "Effect of calcium chloride addition on sulfate resistance of concrete placed and initially cured at 40 and 70 °F", *Concrete Laboratory Report No. C-900*, Bureau of Reclamation, Denver, CO.

Snyder, K.A. and Clifton, J.R. (1995) "4SIGHT: A computer program for modeling degradation of underground low level concrete vaults", *NISTIR 5612*, National Institute of Standards and Technology.

Terzaghi, R.D. (1948) "Concrete deterioration in a shipway", *Journal of the American Concrete Institute* **44**: 977–1005.

van Zeggeren, F. and Storey, S.H. (1970) "The computation of chemical equilibrium", London, Cambridge University Press, UK.

Verbeck, G.J. (1968) "Field and laboratory studies of the sulphate resistance of concrete", in E.G. Swenson (ed.) *Performance of Concrete-Resistance of Concrete to*

*Sulphate and Other Environmental Conditions: A Symposium in Honour of Thobergur Thorvaldson*, University of Toronto Press, pp. 113–125.

Walton, J.C., Plansky, L.E. and Smith, R.W. (1990) "Models for estimation of service life of concrete barriers in low-level radioactive waste disposal", in *Report NUREG/CR-5542*, US Nuclear Regulatory Commission.

# 8   Case histories

Since the introduction of relevant standards and codes in industrialized countries, occurrence of *internal* and *external* sulfate attack in properly designed, processed and executed concrete is rare. When damage occurs, it is always the consequence of incorrect construction that enables penetration into concrete of aqueous salt solutions needed to initiate and feed the attack. Most of the codified recommendations are based on prescription of maximum values for water–cement ratio, maximum levels of $C_3A$ in cement and, in some cases, of minimum cement content and addition of supplementary materials such as selected pozzolanas or slags, or both.

As the sulfate-generated distress is largely a function of concrete quality, the primary objective of the precautionary measures is to decrease the accessibility of sulfate bearing solutions into concrete by decreasing its permeability. A well-constructed, impermeable concrete structure will not suffer from sulfate attack regardless of the prevailing environmental conditions and physico-chemical mechanisms (e.g. potential for ettringite, thaumasite, gypsum, or efflorescence formation). According to Mehta and Monteiro (1993):

> The quality of concrete, specifically a low permeability, is the best protection against sulfate attack. Adequate concrete thickness, high cement content, low water/cement ratio and proper compaction and curing of fresh concrete are among the important factors that contribute to low permeability. In the event of cracking due to drying shrinkage, frost action, corrosion of reinforcement, or other causes, additional safety can be provided by the use of sulfate-resisting cements.[1]

In other words, properly designed and constructed concrete will be stable under most aggressive conditions unless the concentration of sulfates in the soil or the water in contact with the concrete is extreme. Under such conditions additional measures have to be taken to prevent direct contact between the concrete and the $SO_4^{2+}$ source.

However, problems do occur, and sulfate attack may become a real issue when concrete is improperly proportioned, designed, cured and placed in a hostile environment, or both (e.g. Swenson 1968; Mehta 1992; DePuy 1997; Figg 1999). The following case studies are examples that resulted from inadequate utilization of knowledge on concrete mixture design, concrete processing, and its inappropriate use in a potentially hostile environment.

## 8.1   DETERIORATION OF RESIDENTIAL BUILDINGS IN SOUTHERN CALIFORNIA

A well-publicized problem involving residential housing construction in Southern California is an interesting case of *external* sulfate attack (e.g. Reading 1982; Novak and Colville 1989; Rzonca *et al.* 1990; Haynes and O'Neill 1994; Travers 1997; Lichtman *et al.* 1998; Haynes 2000). Numerous court cases were concluded or are still in progress. The alleged violations of best concrete-making practices and codes seem to have lead to premature deterioration of relevant structures, including post-tensioned floor slabs, garage floors, footings, foundations, driveways, retaining walls, and street curbs. The technical explanations of the observed damage, and even the answers to the question whether there is any damage, differ from expert to expert (e.g. see presentations/discussions by Haynes, Diamond and Lee, and others in references Marchand and Skalny (1999) and Haynes (2000)).

Visible changes to concrete were observable often as early as 2–4 years after casting (see Figure 8.1). Structural and other problems unrelated to concrete were also encountered; these will not be highlighted in the following paragraphs.

*Figure 8.1* California residential house footing exposed to sulfate-containing ground waters. Note spalling and efflorescence (Photo: J. Skalny).

It is known for many years, that wide areas of Southern California have soils containing high levels of sulfates, often in form of gypsum (e.g. Novak and Colville 1989; Rzonca *et al*. 1990; Day 1995). Due to the geological history of Southern California – formerly sea beds with heavy salt deposits; earthquake zone – analyses of soil samples revealed variable sulfate concentrations in a wide range from practically nil to well above 10,000 ppm. For this reason, and probably others, the Cement Industry Technical Committee of California issued in the 1970s a "Recommended Practice to Minimize Attack on Concrete by Sulfate Soils and Waters" (CITC 1970). The document clearly states that "low water cement ratio and high density concrete is imperative at all sulfate levels" and recommends the maximum w/cm, minimum cement content, and cement type to be used at various levels of sulfate in ground water. Generally, these recommendations are in line with recommendations or requirements of ACI, Uniform Building Code, California Department of Transportation, and other codes and standards.

The building boom of the 1980s and 1990s led to a situation in which the recommendations with respect to the type of cement were usually followed, but only the bear minimum cement content was used, and the requirements for the maximum allowed w/cm seems to have been often ignored. Excessive w/cm can clearly result in higher concrete porosity and permeability than is appropriate for environment known to have high sulfate concentrations in soil.

As discussed earlier, depending on the concrete quality and environmental conditions, the complex sulfate attack mechanisms may lead to various chemical and physical changes in concrete. *Chemical* changes may include:

1  removal of $Ca^{2+}$ from some of the hydration products (e.g. decomposition of calcium hydroxide and C-S-H, or both);
2  unusual changes in pore solution composition;
3  formation of hydrated silica (silica gel);
4  decomposition of still unhydrated clinker minerals;
5  dissolution of previously formed hydration products;
6  formation of ettringite (in excess of that formed from original sulfate in the cement), gypsum, and thaumasite;
7  formation of magnesium-containing compounds such as magnesium hydroxide (brucite) and magnesium silicate hydrate;
8  repeated recrystallization of sodium sulfate unhydrate (thenardite) to/ from sodium sulfate decahydrate (mirabilite); and
9  penetration into concrete of ionic species and subsequent formation and crystallization of salts such as NaCl, $K_2SO_4$, $MgSO_4$, etc.

The observable *physical* changes are the consequence of the above chemical changes and may include:

1  complete restructuring of the pore structure and solid microstructure;
2  increased porosity and permeability;

3    volumetric expansion and the associated microcracking;
4    formation of complete or partial circumferential rims or gaps (paste expansion cracks) around the aggregate particles;
5    surface spalling, delamination, exfoliation;
6    paste softening, decreased hardness;
7    deposition of salts on surfaces and exfoliation cracks;
8    loss of strength; and
9    decreased modulus of elasticity.

By themselves, neither of the above chemical, physical, and microstructural changes are necessarily an adequate sign of sulfate attack. However, in combination, there can be little doubt. It should be noted that initially, due to pore filling by the reaction products, the reactions of sulfate attack might lead to decreased porosity and even increased compressive strength (e.g. Jambor 1998). However, as the chemical and microstructural changes proceed, the trend reverses and the concrete gradually loses its required engineering properties.

The following information is available from Southern California regarding the relevant conditions and observed phenomena (e.g. Haynes and O'Neill 1994; Day 1995; Deposition Transcripts 1996–2000; Lichtman *et al.* 1998; Diamond and Lee 1999; Brown and Badger 2000; Brown and Doerr 2000; Diamond 2000):

- Large amounts of concrete were designed and placed using w/cm as high as 0.65 (in apparent violation of applicable codes and recommendations); occasionally, concretes with w/c of 0.7 or higher were identified;
- Typical cement content used was about $250–320 \, kg/m^3$ ($400–500 \, lb$ per cubic yard). In some instances, the cement content was as low as $220 \, kg/m^3$ ($350 \, lb$ per cubic yard). Mostly ASTM Type V, in some instances Type II cements with pulverized fly ash were used;
- Compressive strength required at twenty-eight days, depending on the application, was about $13–20 \, MPa$ (*c.* 2,000 to *c.* 3,000 psi);
- Sulfate concentration in ground water is variable, often even within the same construction locality; typically between 150 and 10,000 or more ppm;
- Presence in ground water of $Mg^{2+}$, $Na^+$, $K^+$, $Cl^-$, and $HCO_3^-$ and other ions, in addition to $SO_4^{2-}$;
- Depth of ground water variable from locality to locality, from near-surface to several meters below the surface;
- Typical (summer) ambient temperature: $10–20 \, °C$ at night, $25–35 \, °C$, or more at direct sun exposure, during daytime;
- Humidity variable from very low at daytime to above dew point at night;
- Visually observable damage includes efflorescence, delamination of mortar, exposed aggregate, spalling, and limited cracking;
- Petrographic observations (light optical and SEM microscopy; energy-dispersive spectrometry):[2] formation of ettringite "nests" in the paste, microcracking of the cement paste, expansion of the paste (formation of

circumferential gaps around aggregate particles), formation of gypsum veins, removal of calcium hydroxide from the paste, decomposition of C-S-H, increased and irregular porosity or both, severe carbonation of external and of some buried concrete surfaces, decalcification of the still unhydrated calcium silicates, deposition of reaction products in pores formed as a result of hydration or decalcification of the clinker minerals, formation of Mg-rich layers in Headley grains, formation of brucite and magnesium silicates, presence of Friedel's salt (calcium chloroaluminate hydrate, a chloride analog of calcium monosulfate 12-hydrate), surface efflorescence (predomiantly sodium sulfate; occasionally also sodium chloride, magnesium sulfate, other salts); and some corrosion of reinforcement;

- X-ray diffraction data: presence in the efflorescing material of thenardite or mirabelite; occasionally other salts, including Friedel's salt, NaCl, and $MgSO_4$.
- Physical testing data: decreased hardness of concrete with depth, compressive strengths variable (higher, lower or as designed for twenty-eight days, depending on local conditions and age of concrete exposure to the environment), decreased tensile strength, very high permeability (ASSHTO T 277-831 rapid chloride permeability test, water vapor and water permeability), decreased modulus of elasticity.

In the following, we will present microscopic evidence that has been used in interpretation of some of the observed external sulfate attack phenomena. The set of micrographs in Figure 8.2 is typical of ettringite forms found in concrete exposed to external sulfate attack. As has been discussed earlier, the observed ettringite morphologies found in internal and external sulfate attack situations are similar. This is not surprising considering that the mechanisms are based on the same chemical principles. As emphasized on previous pages, the presence of ettringite *per se* is not a sign of sulfate attack. Ettringite is found usually in the form of:

1   "nests" located throughout the paste in the C-S-H mass, dominating the local morphology and often being accompanied by microcracking and development of a network of microcracks (micrographs a and b);
2   deposits located in gaps around the aggregate particles and in cracks (micrographs c and d);
3   deposits in air voids, usually filling the void only partially (micrograph e); and
4   microcrystalline ettringite, not detectable by microscopic techniques; this form of ettringite is most probably responsible for the paste expansion evidenced by the formation of the gaps around aggregate, and the subsequent deterioration of physical properties.

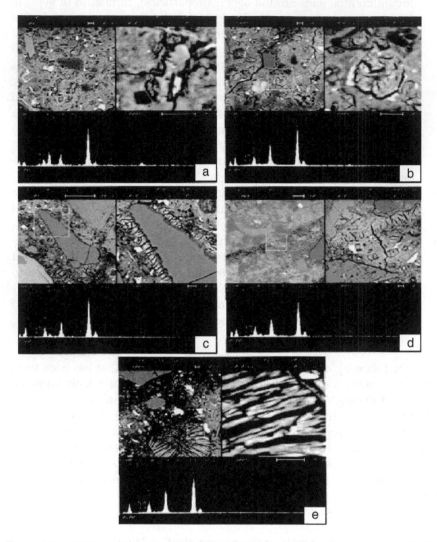

*Figure 8.2* (a,b) Formation of ettringite "nests" and cracking of cement paste; (c,d) ettringite in paste and gaps at the paste-aggregate interface; (e) ettringite partially filling an air void. SEM, backscattered mode (Photos courtesy of RJ Lee Group).

Under specific environmental conditions such as lower ambient temperature and presence of carboxyl ions, thaumasite may form in addition to ettringite. It is believed by some that damage caused by thaumasite may be even more severe that that caused by ettringite. An example of thaumasite crystallites found in concrete exposed to sulfate-containing ground water is given in

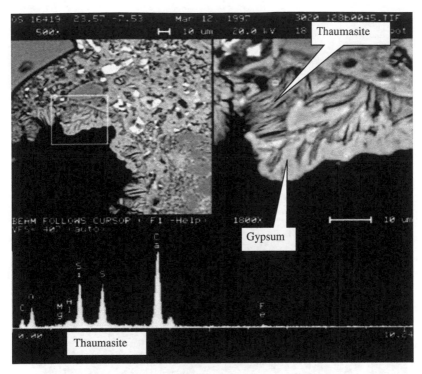

*Figure 8.3* Simultaneous formation of thaumasite (square) and gypsum (triangle) in concrete exposed to external sulfate. EDAX pattern: thaumasite. SEM, backstattered mode (Photo courtesy of RJ Lee Group).

Figure 8.3. Note that this concrete was produced and located in an arid zone where low temperatures, assumed by some to be needed for thaumasite formation, are uncommon.

One of the common observations in sulfate attack is change in paste porosity. Depending on the age of concrete, severity of the sulfate attack and, possibly other variables, the porosity at the time of observation may be unchanged (usually slightly-damaged concrete) or changed dramatically (highly-damaged concrete). Micrographs of Figure 8.4 show typical examples of high (a) and inhomogeneously distributed (b) porosity. The given examples represent concrete made with initial (mix) w/cm of about 0.65.

One of the characteristic features of severe external sulfate attack is formation of gypsum. It is usually found in the form of layered deposits parallel to the surface that is in contact with the sulfate-bearing ground water or soil. Examples of gypsum deposits found in Southern California concrete are shown in Figure 8.5.

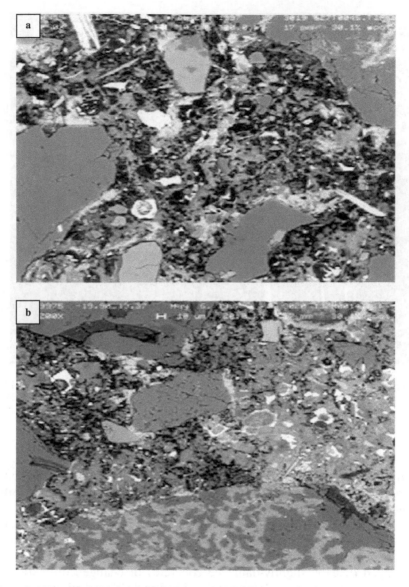

*Figure 8.4* Extremely high (a) and inhomogeneous distribution (b) of porosity. SEM, backscattered mode (Photo courtesy of RJ Lee Group).

In permeable concrete, especially in situations where a part of the above ground concrete is exposed to repeated temperature and humidity fluctuations, the sulfate-bearing solutions may penetrate to the exposed concrete surface where they crystallize. According to ACI Guide to Durable Concrete (ACI 1992), under such conditions concrete may be exposed to severe

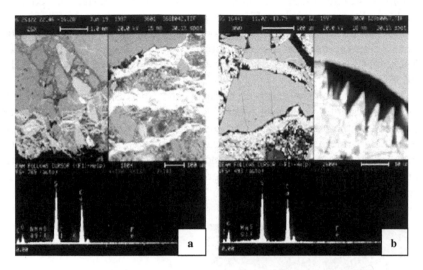

*Figure 8.5* Formation of gypsum veins parallel with the horizontal concrete surface in contact with sulfate-containing ground water. Note well developed gypsum crystals shown on left (Photo courtesy of RJ Lee Group).

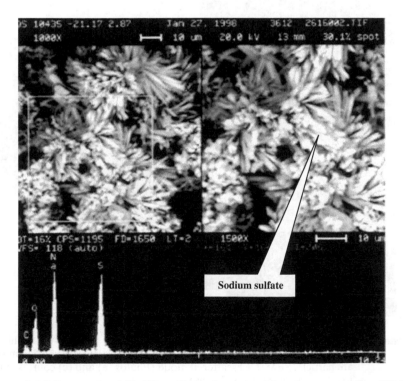

*Figure 8.6* Micrograph of $Na_2SO_4$ efflorescing material and corresponding EDAX patterns. SEM, secondary electron mode (Photo courtesy of RJ Lee Group).

*chemical* sulfate attack. Examples of effloresing material formed under such conditions are given in Figure 8.6.

Deposition of various salts may occur not only at the concrete surface but also in the interior of a concrete structure exposed to ionic solution. Such depositions of NaCl, $Na_2SO_4$, and Friedel's salt are presented in Figure 8.7.

Under conditions where both chloride and sulfate ions are a part of the aggressive solution, one can identify microstructures in which the reaction products of the reinforcement corrosion penetrate, and possibly replace, the localities formerly occupied by hydration products. See Figure 8.8, micrographs a, b. Occasionally, one may encounter evidence of both sulfate attack and reinforcement corrosion (micrograph c).

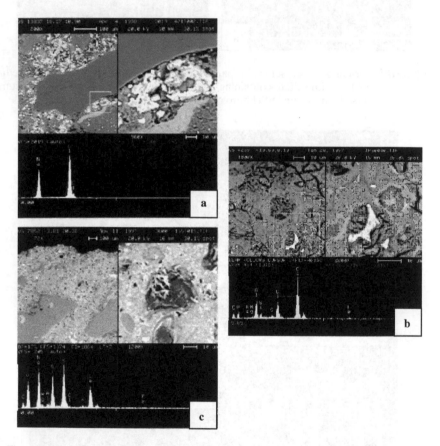

*Figure 8.7* (a) Deposition of sodium chloride in the paste; (b) Friedel's salt within the paste; (c) deposit of sodium sulfate on a partially decalcified calcium silicate particle. SEM, backscattered mode (Photos (a) and (c) courtesy of RJ Lee Group; photo (b): J. Skalny).

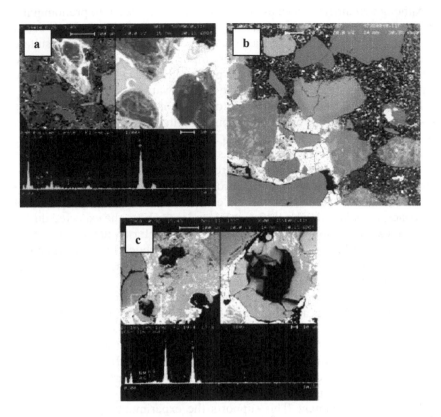

*Figure 8.8* Products of reinforcement corrosion (lightest color). (a,b) Note intermixed paste hydration and corrosion products; (c) penetration of corrosion products into paste and gypsum crystallites. SEM, backscattered mode (Photo courtesy of S. Badger).

Of the "fingerprints" characterizing external sulfate attack, the most convincing in the discussed cases are the presence of gypsum and, if magnesium sulfate is present, of magnesium-containing reaction products, and deterioration of some, though not all, physical properties. Formation of brucite and hydrated magnesium silicates, taking into account the presence of both $SO_4^{2-}$ and $Mg^{2+}$ ions in the ground water, is the most damaging. Their formation is closely associated with decrease in calcium hydroxide concentration and decalcification of the C-S-H. Presence of gypsum in the concrete cannot be explained by any other damage mechanism. It should be noted that of the observed reaction products only brucite and gypsum are stable phases; the other phases present, such as ettringite and monosulfate, are metastable and their presence depends on the local micro-conditions. Gypsum, ettringite and monosulfate can co-exist in cement paste only due to the paste matrix heterogeneity.

Although volumetric changes at macro-scale do not seem to predominate, observation within a few years after casting of micro-scale cracking caused by ettringite formation in the paste is an indication of progressing sulfate attack. As explained above, the lesser importance of the usual ettringite form of sulfate attack is believed to be preconditioned by the environmental conditions and ionic composition of the ground water.

The adequate compressive strength of many of the tested concrete cores has been taken by some as an indication of limited or no damage. However, as is now accepted by most experts (Mehta 1997; Neville 1998; Jambor 1998), strength, especially compressive strength, is an inappropriate measure of durability. Most of the tested concrete had allegedly inadequate tensile and flexural properties, diminished hardness, and lowered modulus of elasticity. The finding that the usual ratio of compressive and tensile properties of the concrete, believed to be *c.* 10:1, increased well above the expected due to microcracks formation may be by itself a sign of internal damage (Ju *et al.* 1999). Such finding was reported earlier (e.g. Harboe 1982). It remains a well-established fact that durable concrete also exhibits adequate mechanical strength, but the reverse may not be the case.

Another reported observation is decomposition of the still unhydrated clinker calcium silicates and their transformation into hydrated (?) magnesium silicate or silica gel. The fact that the original shape of the clinker minerals is maintained indicates that the decalcification happened before these minerals had a chance to hydrate (see micrographs of Figure 8.9). In other words, the aggressive sulfates must have had access to $C_3S$ or $C_2S$ particles in very early stages of hydration, possibly within hours after concrete was placed in the high-sulfate environment. This supports the experimentally obtained data revealing high porosity and permeability.

Damage mechanisms other than external sulfate attack, such as ASR, acid rain, etc., were also considered, but experimental evidence does not support these options. The issue in Southern California seems to be inadequate concrete quality in an environment rich in sulfates and chlorides. Whereas most parties agree that the sulfate levels in California are high and the used w/cm were excessive, the interpretations of the observed damage to concrete vary (e.g. Haynes 2000; Deposition Testimonies 1996–2000). Among others the claim is made that only very few, if any, cases of sulfate attack were documented; those admitted are categorized by some as "physical" salt attack. In contrast, other experts are of the opinion that both chemical sulfate attack and repeated mirabilite–thenardite recrystallization are responsible for the damage, and argue that visual observation is an inadequate technique to assess the sulfate damage to concrete (Diamond and Lee 1999; Deposition Testimonies 1996–2000).

The observed damage to the concrete at macro- and microstructural levels seems to be the result of complex sulfate attack mechanisms involving both chemical and physical processes enabled by high concrete porosity and permeability. Repeated cycles of wetting–drying and of low–high temperatures

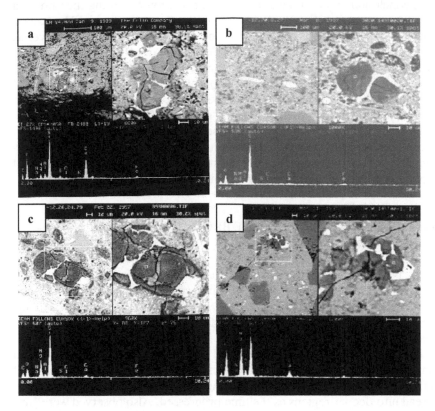

*Figure 8.9* Decalcified pseudomorphs of unhydrated clinker calcium silicates: (a) partial decalcification (Photo courtesy of B. Erlin); (b) complete decalcification to hydrated (?) silica; (c) partial tranformation to magnesium silicate hydrate; (d) complete transformation to magnesium silicate hydrate. SEM, backscattered mode (Photos courtesy of RJ Lee Group).

enabled physical mechanisms to supplement the well-known chemical processes of destruction: formation of internal cracking due to excessive ettringite, decalcification of the paste and C-S-H, and formation of magnesium-bearing compounds.

In addition to the above sulfate-triggered changes, the high porosity of the concrete resulted in excessive carbonation and chlorination. Carbonation is known to affect the stability of ettringite; in carbonated parts of the examined concrete no ettringite was present, and an ettringite layer (or front) ahead of the carbonated zone was observed. The synergistic effect of sulfates and chlorides is not entirely clear, but there are reports on decreased effectiveness of ASTM Type II and Type V cements to sulfates in the presence of chlorides.

In conclusion, we would like to share our opinion regarding the observed damage to concrete in residential houses of Southern California. The issue is not which of the experts are right or wrong. What is unfortunate, however, is that the knowledge on sulfate attack, generated since the 1900s by some of the best "concrete" minds in North America, including California, was somehow forgotten or ignored. All described problems could have been avoided if the most basic principles of concrete making were remembered and adopted. Codes, standards and concrete-making guidelines are clear: sulfate attack can be prevented by production of low-permeability, properly proportioned, and adequately cured concrete. It is the non-compliance with these standards and best practices of concrete making that apparently lead to the encountered problems and subsequent need for expensive rehabilitation.

## 8.2 SULFATE ATTACK DAMAGE BROUGHT ABOUT BY HEAT TREATMENT (DEF)

In the period between 1980 and 1984, an extended occurrence of damage to prefabricated, pre-stressed steel-reinforced concrete railway ties and some other concrete products was observed in what was then West Germany (Association of German Cement Manufacturers 1984). Several millions of ties were affected.

The damage became apparent several years after the products have been manufactured and in use. It was characterized by development of cracks that started at the corners and edges of the concrete element and gradually spread into deeper regions as the time progressed. Also observed were gaps between aggregate particles and the cement paste, associated with a loss of bond between the two. The cracks and especially the gaps tended to be partially or completely filled with crystals of thaumasite alone or a combination of thaumasite and ettringite. There were indications that the cracks and gaps were not created by the formation of thaumasite or ettringite or both, and that these phases only precipitated in already preexisting empty spaces. Remarkably, any damage was observed only in ties that were steam-cured in the course of production and were exposed to rain. No damage was seen in ties located in tunnels or under bridges.

Tests performed on the used cements and aggregates proved that all materials complied with the existing specifications. None of the aggregates was alkali sensitive and alkali–silicate reaction mechanism could be ruled out as the cause of the problem. The strength of the produced concrete exceeded significantly the required value.

In laboratory experiments, triggered by the existing situation in the field, it was found that concrete mixes steam cured at 80 °C exhibited consistently a distinct expansion and even cracking regardless on whether they were stored subsequently at 5 °C or 40 °C, partially immersed in water. The extent of damage was somewhat enhanced in mixes made from cements with elevated

$SO_3$ contents and was reduced in specimens that were precured for three hours at ambient temperature prior to heat curing. The process was always associated with the formation of thaumasite or ettringite or both. Also here, just as in actual use, all these phenomena became apparent only after about three years. The formed cracks were initially empty and became filled with precipitated reaction products as the time progressed. In specimens that were cured at 40 °C rather than at 80 °C, excessive expansion was observed only at elevated $SO_3$ contents in specimens that were not precured.

From the obtained experimental data it was concluded that expansion and cracking of the samples were the result of ettringite formation taking place in the cement paste after steam curing. The different coefficients of expansion of the individual paste constituents were believed to contribute to the formation of the gaps around the aggregate particles. This phenomenon becomes enhanced if an appropriate precuring at ambient temperature is omitted, due to the inability of the mix to develop sufficient strength before it becomes exposed to heating. It was also postulated that steam curing immediately after mixing prevents or reduces the initial formation of ettringite in a reaction between $C_3A$ and calcium sulfate taking place at ambient temperature, which enhances the formation of this phase after the steam curing had been completed.

Based on the field and laboratory experience, the existing problems were eliminated by using cements with moderate $SO_3$ contents, by introducing a minimum precuring period in the production process, and by reducing the maximum steam-curing temperature to below 65 °C (Deutscher Ausschuss fur Stahlbeton 1989; Skalny and Locher 1999).

## 8.3 CONCRETE RAILROAD SLEEPERS: HEAT-INDUCED INTERNAL SULFATE ATTACK (DEF) OR ASR?

As is the case with external sulfate attack, *internal* sulfate attack, caused by presence or renewed availability of reactive sulfate in concrete components, is rare. The excess of sulfate may originate from the cement (sulfate in excess of that allowed by standards), from aggregate (e.g. oxidation of pyrite, presence of gypsum), or from other materials added to concrete.

However, as discussed in more detail earlier in Chapter 4, sulfate may also become available as a result of improper concrete processing – particularly in cases involving improper curing. An example of such attack is the *DEF*-form of internal sulfate attack.

The to be described, well-publicized case in the United States, involving concrete railroad ties, is based on open literature and other available documents (e.g. Trial Transcripts 1993–1994; Mielenz *et al.* 1995; Johansen *et al.* 1993; Federal Supplement 1995; Scrivener 1996). It is technically not a unique case, as other similar cases were documented in recent past in Germany (see above Case Study), Canada, Finland, ex-Czechoslovakia, South African Republic, Sweden (Rise 2000), and several other countries world over.

About half a million of concrete railroad sleepers were produced in a precast operation between 1983 and 1991. The aggregate as well as a limited portion of the ASTM Type III cement (cement A) used in the production of the ties, have been produced by the company that owned the concrete precast plant. The majority of the Type III cement used was produced by another cement company (cement B). The two cements differed somewhat in total alkali content (B higher than A) and a limited proportion of cement B was alleged to have $SO_3$ content somewhat outside the ASTM specification. The aggregate was predominantly deformed granitic rock abundant in microcrystalline quartz (Kerrick and Hooton 1992). Such aggregates are known to be alkali reactive.

Type III cement in conjunction with steam curing was used on a 24-hour cycle to guarantee adequate strength at demolding time. Some of the produced ties were not steam-cured, but were cured under ambient conditions over the weekend; such ties were referred to as "Friday ties." The plant operated four casting beds, each of which consisted of seventy forms holding eight ties each (a total of $70 \times 8 = 560$ ties). Towards the end of the working day, a w/cm = 0.42 concrete mix was placed and vibrated in the forms. Upon precuring, the whole system was exposed to live steam overnight. In the morning the ties were demolded and subsequently stored in an outdoor storage area. Each tie was imprinted with identification of pour number and form number, allowing identification of production parameters.

Within a few years, a substantial proportion of the ties developed visible longitudinal cracks between the rail seatings and map cracks at the tie shoulders, such as shown in Figures 1.3 and 8.10. This led to a complex litigation and still-persisting discussions in the technical community. The questions asked were:

- What was the primary mechanism of deterioration: alkali–silica reaction (ASR) or so-called delayed ettringite formation (DEF)? Were there other damage mechanism operable?
- Who is responsible for the damage: the precast concrete producer or the cement producer?

For the purpose of this discussion, it is of no consequence which party was correct; the issue of interest is better understanding of the observed phenomena and prevention of similar concrete damage in the future.

Based on substantial amount of theoretical and experimental work, on analysis of the production practices in the concrete precast plant, and on analysis of the frequency and level of damage to ties, the following can be concluded:

1   On a scale of 1 (no cracks) to 5 (well-developed map cracks; some cracks over 1 mm wide; some pre-stressing wire pull-back), about 8% of the reviewed tie population exhibited severe cracking (ratings 4 and 5).
2   Petrographic analysis clearly demonstrated that the aggregate used in the tie production was alkali–silica reactive (Kerrick and Hooton 1992).

*Figure 8.10* Concrete railroad tiles showing map cracking at shoulders and longitudinal cracks between rail seatings (Photos courtesy of J. Skalny and N. Thaulow).

The severity of ASR increased with increasing tie rating. In the most damaged ties (rated 5), the majority of the aggregate particles had been cracked and ASR reaction products were associated with these cracks (see Figure 8.11).

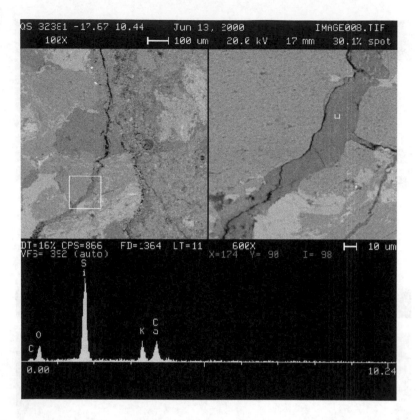

*Figure 8.11* Aggregate crack caused by formation of ASR gel (Photo courtesy of S. Badger).

3   Secondary ettringite was found in both undamaged and damaged ties, predominantly in pores (present in all ties) and in cracks associated with ASR. Such ettringite is believed to be formed by recrystallization through solution and is not believed to be the cause of expansion.

4   In a portion of the examined ties, an ASR-independent phenomenon was noticed, namely formation of circumferential gaps around the aggregate particles (see ettringite veins in Figure 8.12; compare with Figure 4.4). These gaps were usually, though not always, partially or completely filled with ettringite. The incidence of these gaps was linked to those ties that were steam cured, particularly to those that were during steam curing located above the ends of the pipes carrying the live steam to the curing bed. Notwithstanding this form of severe damage, these ties were also severely affected by ASR. This feature was observed in ties made with both cements A and B used during the production period at issue, as well as in ties made with a third cement.

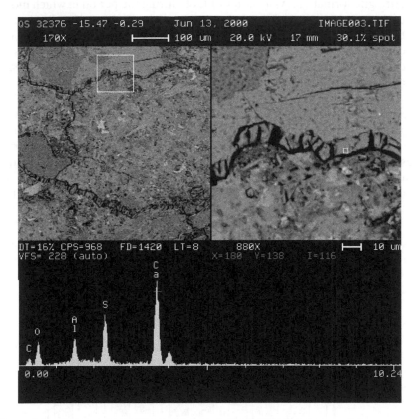

*Figure 8.12* Micrograph showing secondary ettringite-filled gaps around the aggregate particles and in paste cracks in a tie rated 5 (Courtesy of N. Thaulow).

5   Analysis of the production practices and a survey of ties showed that 5-rated ties were almost entirely absent from non-steam cured "Friday ties" produced during weekends. Data on the exact temperature of concrete were not readily available, but excessive curing temperatures were suspected. The incidence of 5-rated ties from steam-cured pours was significantly higher at the positions above the ends of the pipes carrying steam to the beds. Ties produced on the same day with the same cement (whether A or B) developed very different degrees of damage in the field depending on their position in the casting bed. Microscopic examinations highlighted earlier and analysis of the production data (Figure 8.13) are consistent in showing the effect of production processes on the deterioration of the ties, independent of the cement used.

6   Although suspected, there was no experimental evidence of any abnormal sulfate form in any of the cements or ties examined. No linkage could be made between high clinker sulfate levels and degree of damage due to

ettringite formation. As a matter of fact, during the period in which most of the severely damaged ties were produced, the clinker sulfate levels are believed to have been lower than usual. Please, note that expansion of concrete subjected to elevated temperatures and associated with so-called DEF can occur in most Portland cement concrete; it did occur in all three cements discussed above.

From the above data it is obvious, and most literature agrees, that ASR was the primary mechanism of cracking of the ties, and occurred in ties made with both cements. Exposure of some of the concrete ties to excessive temperatures was a secondary cause of cracking and, again, it occurred in ties made from both cements. It is almost certain in our view, that in the absence of primary damage due to ASR, the ties would not have been damaged to the

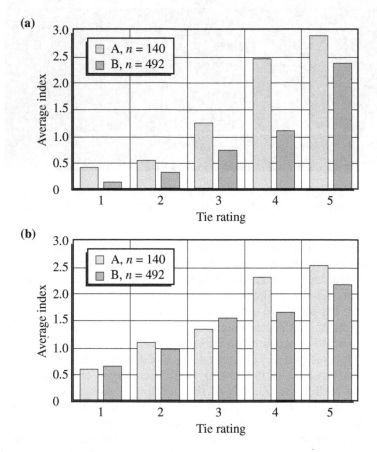

*Figure 8.13* Ettringite (a) and ASR (b) in concrete ties made with cements A and B as a function of tie rating.

observed degree, if at all. This interpretation is fully supported by recently published data by Taylor (1997), Tennis *et al.* (1997), Thomas (1998, 2000) and others. Neither the alleged negative effects of clinker sulfate form and amount, nor the possibility of low-temperature DEF were experimentally substantiated as of today.

Needless to say that low-temperature sulfate attack is possible in cases such as over-sulfated cement, sulfates in aggregate, etc. As discussed earlier, the actual mechanisms of internal sulfate attack or the differences, if any, between the mechanisms of thermally-induced (DEF) versus excess sulfate-induced ("classical" internal) attacks are not entirely clear at the present time. However, it remains a fact that ettringite expansion is possible only when the conditions allow *primary* ettringite crystallization; *secondary* ettringite, meaning ettringite formed by dissolution and recrystallization or growth from solution, is a non-expansive process. Thus, the observance of ettringite in pores, cracks or voids in concrete is an inadequate proof of "delayed" ettringite formation.

Not everybody's opinions are in line with all of the above conclusions (e.g. Heinz and Ludwig 1987; Mielenz *et al.* 1995). However, the majority of the scientific community agrees today that (a) heat-induces sulfate attack is more prevalent in the presence of ASR; (b) there is no credible experimental evidence for "ambient-temperature DEF"; and (c) more data are needed to elucidate the complex relationship between the cement sulfate levels, the temperature of curing, and other less-recognized variables.

## 8.4 DETERIORATION OF UK CONCRETE BRIDGE FOUNDATIONS CAUSED BY THE THAUMASITE FORM OF SULFATE ATTACK[3] (TSA)

*Introduction*   In March 1998, Building Research Establishment (BRE) were commissioned by the United Kingdom Highways Agency (HA) to establish the degradation process responsible for surface damage to buried columns and base slabs of three piers of a motorway overbridge. Halcrow Materials Specialists identified the problem, following core sampling and testing in February 1998, as likely due to the thaumasite form of sulfate attack (TSA). This was confirmed by BRE, who found it had been due to exposure of the buried bridge foundations to sulfates present in the adjacent re-worked Lower Lias Clay (of Jurassic age) that had been used as backfill in the foundation excavations. The bridge in question, the 30-year-old Tredington–Ashchurch bridge, is situated just south of Junction 9 on the M5 in Gloucestershire in the West of England. The deterioration, which manifested itself as a softening of the outer surface of the concrete, was noticed by engineers working for Halcrow Group Ltd (Halcrow) during routine strengthening of the bridge piers. BRE carried out a site investigation followed by laboratory studies on samples of the surrounding ground and on concrete samples taken from the affected bridge foundations. The results are presented in this brief case study. More in-depth accounts of the Tredington–Ashchurch case study have been reported elsewhere (TEG Report 1999; Crammond *et al.* 2001; French 1999; Halcrow Report 2000a, 2000b).

*Description of bridge.*   The bridge was constructed during 1968–1969 and prior to 1998, when exposed for routine bridge strengthening, the concrete below ground had never been excavated and examined. Each of the three bridge piers comprised three slender reinforced concrete columns (0.46 m × 0.76 m in section and 9.1 m in height) founded on a reinforced concrete base slab (4.1 m × 13.3 m in plan by 0.91 m deep).The base slabs had been constructed in excavations sunk about 3.7 m below original ground level, and then, after construction of the columns, backfilled with the locally derived Lower Lias Clay material. The motorway embankment (also of Lower Lias Clay) and road pavement were then constructed over this backfill, covering the original ground surface to a depth of 2 m. The bridge was cast *in situ*, but the upper surface of the base slab and the buried sections of the columns would have been exposed to air for a period of time before burial. These surfaces could therefore have become carbonated. In 1998, at the time of bridge strengthening, the columns to all three piers were excavated below ground level down to the top of the base slab level in pits on the hard shoulders and in the central reserve.

*Ground conditions contributing to TSA.*   A study of the surrounding ground was carried out during the bridge site investigation in order to identify characteristics and processes at work, which might have contributed to the TSA. The affected concrete foundation elements of the Tredington–Ashchurch bridge were in contact with a large volume of re-worked, initially unweathered, pyritic Lower Lias Clay. The principal conclusion is that sulfate in the relatively large quantity of Lower Lias Clay backfill around the foundations provided the source of external sulfates needed to fuel the TSA.

The process of sulfate formation in Lower Lias Clay is as follows. If oxygen-rich ground water has access to unweathered clay it reacts with the finely disseminated pyrite ($FeS_2$) to form red-brown ferric oxide ($Fe_2O_3$) together with sulfuric acid ($H_2SO_4$). The latter further reacts with calcium carbonate ($CaCO_3$), which is abundant in the Lower Lias Clay, to produce calcium sulfate which crystallizes as gypsum ($CaSO_4 \cdot 2H_2O$). Under natural conditions this oxidation process is extremely slow, taking of the order of thousands of years to penetrate several metres below ground surface. However, this process was much more rapid in the case of the Tredington–Ashchurch Bridge, as mechanical reworking exposed and aerated the clay prior to its use as backfill. Sulfide oxidation of the reworked clay could also have been accelerated by bacterial action, in particular the bacterium Thiobacillus ferrooxidans. The presence of this extra sulfate in the clay and ground water increased the sulfate conditions from Class 1 (for the initially undisturbed and unweathered Lower Lias Clay) to Class 3 and above in accordance, respectively, with the UK specifications at the time of construction (Digest 90 1968) and those current in 1998 (BRE Digest 363 1995).

Ground water conditions around the TSA-affected bridge foundations were highly adverse. Sulfate-bearing ground water was ponded around the concrete within a sump formed by the backfilled construction excavation. The ground water in the sump was replenished by leakage from perforated drainage pipes, which passed through the top of the backfill.

Bearing in mind that aggressive TSA generally requires temperatures of below 20 °C, it is perhaps also significant to note that, unlike most building foundations, there was no source of heat at the Tredington–Ashchurch bridge site which could have raised ground temperatures above natural levels and possibly have retarded the TSA process.

*TSA in concrete of bridge piers.*   Surface deterioration due to TSA was present mainly at the base of the buried columns of the Tredington–Ashchurch Bridge up to a height of about three metres above the foundation base slab as shown in Figure 8.14. The TSA had resulted in severe softening and expansion of the outer surface of the affected concrete. In these locations, the surface of the concrete was coated with a white friable material, comprised mostly of thaumasite. This material was easily removed with tools or even by hand. The depth of the TSA into the concrete varied but had reached 30 mm or more in places. As the depth of the embedded steel reinforcement in the columns was around 30 mm, reinforcement corrosion had started in some areas where the cover had been severely damaged. The corrosion may have been aggravated by high chloride levels in the ground water derived from motorway de-icing salts.

Four cores (two 75 mm and two 100 mm) and two lump samples were taken from the bridge foundations for examination and analysis at BRE. The techniques used included X-ray Diffraction, Optical Microscopy, Scanning Electron Microscopy and Chemical Analysis. It was found that further into the columns and base slab beyond the deteriorated zone, the bulk of the concrete was of a high quality with low porosity, good compaction, an average cement content of 370 kg/m$^3$ and a water–cement ratio of about 0.5. The cement type used was identified as ordinary Portland cement (OPC). Good quality natural aggregates were used, namely an unwashed dolomitic, Carboniferous limestone coarse aggregate, and a local fine aggregate containing oolitic limestone (6 mm to 2.36 mm) and a fine quartz sand (<2.36 mm). According to the specifications current at the time of construction (Digest 90 1968), the concrete used in the columns and base slab should have been sulfate resistant and fit for purpose for the then assessed ground conditions. It would also have been sulfate resistant in Class 2, but not in Class 3 conditions.

The limited results obtained from the BRE study indicate that the level of sulfates quickly subsides beyond the point where thaumasite is no longer detected petrographically. This means that sulfate ions did not penetrate into the concrete to any significant depth beyond the degraded surface material.

*TSA reaction mechanism.*   In addition to sulfate ions, TSA requires two further ingredients and these are calcium carbonate ($CaCO_3$) and calcium

*Figure 8.14* Extreme example of TSA in a Tredington–Ashchurch Bridge column
exposed to wet, reworked Lower Lias Clay.

silicate cementitious phases. In the concretes examined from the Tredington–
Ashchurch bridge, the calcium carbonate was derived from four possible
sources: the dusty material around the dolomite coarse aggregate; as a result
of de-dolomitization of the dolomite; from the oolitic limestone finer aggregate; and from carbonates dissolved in the ground water (French 1999). The

calcium silicate was derived from the Portland cement paste both in the form of hydrates and remnant clinker grains.

The TSA reaction sequence was found to consist of three distinct stages as follows:

*Stage 1*  Cracking develops within the cement paste matrix and this runs sub-parallel to the surface of the concrete. Several such cracks are formed, which are very fine to start with but gradually open up as they are filled with thaumasite. The thaumasite crystals grow so that their long acicular axes run perpendicular to the crack walls. The action of these cracks being filled up with a secondary reaction product is obviously an expansive process.

*Stage 2*  The existing thaumasite-filled cracks become progressively wider and haloes of thaumasite soon begin to form around aggregate pieces, especially those composed of limestone (Figure 8.15). Once again, this is an expansive process.

*Stage 3*  Eventually, the whole cement paste matrix disappears leaving isolated aggregate particles embedded in a mass of thaumasite. This stage of the deterioration process was only observed in lump samples A and B and not in the core samples implying that it can be destroyed and lost during coring on site.

*Figure 8.15*  Hand specimen of deteriorated Tredington–Ashchurch Bridge foundation concrete showing the white haloes of thaumasite around the dolomite coarse aggregate particles.

The thaumasite formed during all three stages was shown by micro-analysis to have a very consistent and pure composition, but its crystalline status was seen to vary greatly from practically amorphous to fully developed.

The three stages of TSA development quoted here vary slightly from the four zones quoted in the TEG Report (1999). The reason for this is that, in subsequent investigations of a large number of specimens by others (French 1999; Halcrow Report 2000a, 2000b), some thaumasite and secondary ettringite precipitation was found in the voids of the relatively sound concrete immediately behind the TSA-affected outer zones.

*Bridge refurbishment.* The three slender columns of each pier were completely removed during refurbishment of the bridge and were replaced with a more robust construction comprising much thicker sections of TSA-resisting concrete. Re-worked Lower Lias Clay was not used in the reconstruction.

## ACKNOWLEDGMENTS

The topics described in this case study comprise part of research by BRE into the thaumasite form of sulfate attack. This research is sponsored by the Department of Environment Transport and the Regions. The site work was carried out at the invitation of the Highways Agency and in co-operation with Halcrow, engineering consultants working at that time for Gloucestershire County Council as agents for HA.

## NOTES

1 Use of sulfate-resisting cements is considered to be a *secondary* protection supplementing good concrete quality!
2 It is important to note that the use of electron-optical techniques in forensic evaluation of concrete is relatively new, thus still considered controversial and unreliable by some. However, in the authors' opinion, proper combined use of scanning electron microscopy (SEM) with energy-dispersive X-ray analysis (EDAX) and other techniques enables qualitative and semi-quantitative determinations in most cases. Confrontation of such data with those of optical microscopy, X-ray diffraction, and other techniques is recommended.
3 Prepared by N.J. Crammond, T.I. Longworth, and R.G. Sibbick.

## REFERENCES

ACI (1982) *Goerge Verbeck Symposium on Sulfate Resistance of Concrete, ACI Publication SP-77*, 7 papers, 94 pp.
ACI (1992) *Guide to Durable Concrete*, ACI 201.2R-92, p. 10.

Association of German Cement Manufacturers (Verrein Deutscher Zementwerke) and Research Institute of the Cement Industry (Forschunsinstitut der Zement-industrie) (1984) Activity Report 1981–1984 (in German).

BRE Digest 363 (1995) "Sulfate and acid resistance of concrete in the ground", *Construction Research Communications*, BRE.

Brown, P.W. and Badger, S. (2000) "The distribution of bound sulfates and chlorides in concrete subjected to mixed NaCl, $MgSO_4$, $Na_2SO_4$ attack", *Cement and Concrete Research* **30**: 1535–1542.

Brown, P.W. and Doerr, A. (2000) "Chemical changes in concrete due to the ingress of aggressive species", *Cement and Concrete Research* **30**: 411–418; see also relevant discussion by W.G. Hime and S. Marusin and response by P.W. Brown and A. Doerr (2001) *Cement and Concrete Research* **31**: 157, 159–160.

CITC (1970) Cement Industry Technical Committee of California, *Recommended Practice to Minimize Attack on Concrete by Sulfate Soils and Waters*.

Crammond, N.J., Longworth, T.I. and Sibbick, R.G. (2001) "Tredington–Ashchurch Bridge on the M5: Examination of TSA-affected concrete foundations and surrounding ground", to be published as a *Special Report by Construction Research Communications Ltd*, UK.

Day, R.W. (1995) "Damage to concrete flatwork from sulfate attack", *J. of Performance of Constructed Facilities* (ASCE), pp. 302–310.

Deposition Transcripts (1996–2000) Deposition Testimonies of Messrs. P.W. Brown, S. Diamond, B. Erlin, W. Hime, A. Neville, N. Thaulow, etc.

DePuy, G.W. (1997) "Review of sulfate attack in US and Canada", Report submitted on behalf of plaintiffs in *Murphy v. First Southwest Diversified Partners*, Case #752593 (Orange County Superior Court, California, 1997), 3 November, 89 pp.

Deutscher Ausschuss fur Stahlbeton (1989) *Recommendation on the Heat Treatment of Concrete* (in German), September.

Diamond, S. (2000) "Microscopic features of ground water-induced sulfate attack in highly permeable concretes", presented at the ACI/CANMET mtg., Barcelona, Spain, June.

Diamond, S. and Lee, R.J. (1999) "Microstructural alterations associated with sulfate attack in permeable concretes", in J. Marchand and J. Skalny (eds) *Materials Science of Concrete Special Volume: Sulfate Attack Mechanisms*, The American Ceramic Society, Westerville, OH, pp. 123–173.

Digest 90 (1968) "Concrete in sulfate-bearing soils and ground waters", HMSO, UK.

Federal Supplement (1995) Lone Star Industries Inc. Concrete Railroad Cross Ties Litigation, Federal Supplement 822 (D.Md. 1995).

Figg, J. (1999) "Field studies of sulfate attack on concrete", in J. Marchand and J. Skalny (eds) *Materials Science of Concrete Special Volume: Sulfate Attack Mechanisms*, The American Ceramic Society, Westerville, OH, pp. 315–324.

French, W.J. (1999) "The mechanism of thaumasite sulfate attack", Geomaterials Research Services Ltd, Report Reference 4570B, 13 March.

Halcrow Report (2000a) "Thaumasite investigation, Tredington–Ashchurch road bridge factual report", July 2000, prepared on behalf of The Highways Agency.

Halcrow Report (2000b) "Thaumasite investigation, final interpretative report, vols 1–3", June 2000, prepared on behalf of The Highways Agency.

Harboe, E.M. (1982) "Longtime studies and field experiences with sulfate attack", in *Sulfate Resistance of Concrete* (George Verbeck Symposium), ACI SP-77, pp.1–20.

Haynes, H. (2000) "Sulfate attack on concrete: laboratory versus field experience", *Suppl. Proc. 5th CANMET/ACI Int. Conf. Durability of Concrete*, Barcelona, June (in press).

Haynes, H. and O'Niell (1994) "Deterioration of concrete from salt crystallization, in P.K. Mehta symposium on durability of concrete", *Proc. 3rd CANMET/ACI Int. Conf. Durability of Concrete*, Nice, France, May 1994; see also: Haynes, O'Neill and Mehta, "Concrete deterioration from physical attack by salts", *Concrete International*, January, pp. 63–68.

Heinz, D. and Ludwig, U. (1987) "Mechanism of secondary ettringite formation in mortars and concretes subjected to heat treatment", in *ACI SP 100*, **2**: 2059–2071.

Jambor, J. (1998) "Sulfate corrosion of concrete", unpublished manuscript summarizing his views on sulfate durability of concrete. (Dr Jambor passed away in May 1998.)

Johansen, V., Thaulow, N. and Skalny, J. (1993) "Simultaneous presence of alkali–silica gel and ettringite in concrete", *Adv. Cem. Res.* **5**(17): 23–29.

Ju, J.W., Weng, L.S., Mindess, S. and Boyd, A.J. (1999) "Damage assessment and service life prediction of concrete subject to sulfate attack", in J. Marchand and J. Skalny (eds), *Materials Science of Concrete Special Volume: Sulfate Attack Mechanisms*, The American Ceramic Society, Westerville, OH, pp. 265–282.

Kerrick, D.M. and Hooton, R.D. (1992) "ASR of concrete aggregate quarried from a fault zone: results and petrographic interpretation of accelerated mortar bar tests", *Cement and Concrete Research* **22**: 949–960.

Lichtman, G., Johnson, R. and Steussy D. (1998) "Construction defects underfoot", *Claims*, June, pp. 40–47.

Marchand, J. and Skalny, J. (eds) (1999) *Material Science of Concrete Sperial Volume: Sulfate Attack Mechanisms*, The American Ceramic Society, Westernille, OH, 371 pp.

Mehta P.K. (1992) "Sulfate attack on concrete – a critical review" in J. Skalny (ed.), *Materials Science of Concrete*, vol. III, The American Ceramic Society, Westerville, OH, pp. 105–130.

Mehta, P.K. (1997) "Durability – Critical issues for the future", *Concrete International* **19**(7): 27.

Mehta, P.M. and Monteiro, P.J.M. (1993) *Concrete*, 2nd edn, McGraw-Hill, 548 pp.

Mielenz, R.O., Marusin, S.L., Hime, W.G. and Jugovic, Z.T. (1995) "Investigation of Prestressed Concrete Railway Tie Distress", *Concrete International* **17**: 62–68.

Neville A. (1997) *Properties of Concrete*, 4th edn, Pitman.

Neville, A. (1998) "A 'New' look at high-alumina cement", *Concrete International* **20**(8): 51.

Novak, G.A. and Colville, A.A. (1989) "Effloresence mineral assemblages associated with cracked and degraded residential concrete foundation in Southern California", *Cem. Concr. Res.* **19**(1): 1–6.

Reading, T.J. (1982) "Physical aspects of sodium sulfate attack on concrete", in *ACI SP-77*, American Concrete Institute, pp. 75–81.

Rise, G. (2000) "Deteriorated sleepers – possible mechanisms", Strangbetong, Sweden, August, 25 pp.

Rzonca, G.F., Pride, R.M. and Colin, D. (1990) "Concrete deterioration in east Los Angeles county area: case study", *J. of Performance of Constructed Facilities* (ASCE), 24–29.

Scrivener, K.L. (1996) "Delayed ettringite formation and concrete railroad ties", in *Proc. 18th Int. Conf. Cement Microscopy*, ICMA, pp. 375–377.

Skalny, J. and Locher, F. (1999) "Curing practices and delayed ettringite formation – the European experience", *Cement, Concrete, Aggregate*, ASTM, June.

Swenson, E.G. (ed.) (1968) *Performance of Concrete: Resistance of Concrete to Sulphate and other Environmental Conditions*, A Symposium in Honour of Thorbergur Thorvaldson, University of Toronto Press, 13 contributions, 243 pp.

Taylor, H.F.W. (1994) "Sulfate reactions in concrete – microstructural and chemical aspects", in E. Gartner and H. Uchikawa (eds) *Cement Technology*, The American Ceramic Society, Westerville, OH, pp. 61–78.

Taylor (1996) "Ettringite in cement paste and concrete", in *Beton: du materiau a la structure*, Proceedings, Conference in honor of Micheline Moranville-Regourd, Arles, France, September.

Taylor H.W.F. (1997) *Cement Chemistry*, 2nd edn, Thomas Telford, London.

TEG Report (1999) "The thaumasite form of sulfate attack: risks, diagnosis, remedial works and guidance on new construction", Report of the Thaumasite Expert Group, Department of the Environment, Transport and the Regions.

Tennis, P.D., Bhattacharja, S., Klemm, W.A. and Miller, F.M. (1997) "Assessing the distribution of sulfate in portland cement and clinker and its influence on expansion in mortar", presented at the *ASTM Symposium on Internal Sulfate Attack on Cementitious Systems: Implications for Standards Development*, San Diego, December.

Thomas, M.D.A. (1999) "Delayed ettringite formation in concrete: recent developments and future directions", University of Toronto, 1998; to be published in *Materials Science of Concrete*, vol. VI, The American Ceramic Society, Westerville, OH, 2001, (in press).

Thomas, M.D.A. (2000) "Ambient temperature DEF: weighing the evidence", presented at the ACI/CANMET mtg. in Barcelona, Spain, June.

Travers, R. (1997) "Preventing and defending their (claimed) attack on concrete", *Builder and Developer*, September–October, p. 14.

Trial Transcripts (1993–1994) Lone Star Industries Inc. versus Lafarge Corporation; Trial Testimonies of Messrs. R. Carrasquillo, W. Hime, R.J. Lee, R. Mielenz, L. Spellman, H.F.W. Taylor and N. Thaulow, others.

# 9 Assessment of cement and concrete performance under sulfate attack

The assessment of the expected performance of cement and concrete to be exposed to sulfates may be based on tests or predictions or both. The purpose of assessment is to avoid concrete damage or shortened life due to the use of an inappropriate binder or concrete mixture. An ideal test should be as simple as possible, yield results within a short period of time, and present reliable information about the concrete performance that may be expected under field conditions.

Basically, for convenience, two different types of sulfate attack may be defined, with the testing procedure and method of evaluation of the test results being different for each:

- In *internal sulfate attack*, the deleterious action of sulfates is brought about by an excessive $SO_3$ content in the binder or, less often, in the aggregate used. Under these conditions the whole volume of the material is affected more-or-less uniformly. The extent of damage depends on the composition of the mixture, the curing conditions, and the environment to which the object of concern is exposed.
- In *external sulfate attack*, the sulfates responsible for the damage migrate into the concrete from an outside source. Under this condition, an altered layer resulting from the action of sulfates develops on the surface in contact with the sulfate-containing water or soil, while as the material in deeper regions stays unaffected. The performance of the concrete will depend not only on the binder employed, but to a great degree also on the mixture proportions and the resultant permeability to the sulfate solution.

The following facts have to be considered in assessing the expected performance under sulfate attack:

- A cement may perform differently if exposed to different forms of sulfate (e.g. alkali sulfates, magnesium sulfate, sulfuric acid, etc.).

- The results obtained will depend not only on the binder employed, but also on the mixture proportions, details of concrete processing, and conditions of exposure to the sulfate solution. Thus, standardization of these parameters is necessary in tests aimed at assessing the sulfate resistance of cements.
- The extent of the damage is time-dependent and prolonged curing times may be necessary to make a reliable estimation of the expected performance under field conditions.
- To obtain results within a reasonable time, accelerated testing may be done. However, caution has to be employed if doing so, as the effects obtained may be distorted under these conditions.

The damage from sulfate attack is in most, but not all, instances related to the formation of ettringite (AFt-phase), with a resulting expansion of the cement paste and deleterious effects associated with such an expansion. However, degradation of the C-S-H phase may also take place and, in the case of magnesium sulfate, this effect may mask the sulfate attack. Also, a combined sulfate–acid attack may occur if the substance responsible for the attack is sulfuric acid or ammonium sulfate. All this has to be taken into consideration when designing a *laboratory* testing procedure aimed at assessing the expected effects of sulfate attack under *field* conditions.

Numerous specifications have been developed by different national organizations. Rather than making a comprehensive survey of national standards, in this chapter we will focus on the fundamentals of the problem, but still will discuss some standards, particularly those of ASTM. (ASTM C 150, C 1157, C 1012, E 632; BRE Digest 363 (1996)).

Based on existing experience and long-term tests, *prescriptive standards* have been developed, which specify the properties of the binder or the mortar–concrete mixtures to be used in environments of different aggressiveness, without the necessity to assess the expected deleterious effect in separate tests. In this instance the testing may be limited to checking the criteria set by these specifications, such as the composition of the binder being evaluated. (e.g. Hobbs 1994; Hooton 1999; Patzias 1987, 1991). It has been argued that prescriptive specifications may pose barriers to innovation, as they put rigid limits on the range of permitted composition, however, they serve well, if used for assessing already widely-used types of cements. There is no doubt that the reliability of such specifications improves as they are increasingly based on scientific considerations, rather than just on empiricism.

Contrary to prescriptive standards, *performance standards* describe standardized procedures for a direct assessment of the performance of the cement or concrete under sulfate attack. For example, test specimens of a standard composition, shape, and size may be exposed to sulfate-containing water in a standardized way and the changes that have taken place in a specified time evaluated.

## 9.1　PRESCRIPTIVE STANDARDS FOR ASSESSING SULFATE RESISTANCE OF CEMENTS

Prescriptive standards for assessing the resistance of a Portland cement to external sulfate attack, usually limit the amount of tricalcium aluminate ($C_3A$) in the binder. This compound or its reaction product, monosulfate (Afm), are responsible for poor performance in a sulfate environment. By reacting with sulfate ions, they yield ettringite. In some specifications, the amount of calcium aluminate ferrite ($C_4AF$) in a Portland cement is also limited, as this phase may react in a similar way, though at a reduced rate, and though its expansiveness is distinctly lower. The amount of these phases does not need to be determined directly, but may be calculated from the oxide composition of the cement by the Bogue method. Here, it is assumed that all $Fe_2O_3$ is present in the clinker as $C_4AF$, and $Al_2O_3$ that is not bound in this phase is present as $C_3A$. The particular formulas are as follows:

$$C_3A = 2.6504\ Al_2O_3 - 1.6920\ Fe_2O_3$$
$$C_4AF = 3.0432\ Fe_2O_3$$

where $Al_2O_3$ and $Fe_2O_3$ are the amounts of these oxides in the cement in mass per cent.

The ASTM C150-95 specification for a Type II Portland cement (moderate sulfate resistance) limits the $C_3A$ content to a maximum of 8%. For an ASTM Type V Portland cement (high sulfate resistance), the $C_3A$ content is limited to 5% and the ($C_3A + C_4AF$) content to a maximum of 20%. The permitted $C_3A$ content in the British Standard BS 4027: 1996 (sulfate-resisting Portland cement) is a maximum of 5%.

Unlike Portland cements, the expected performance of blended cements is difficult to assess by prescriptive standards, due to the large variety of factors involved. A direct testing of the binder by the use of a performance test appears necessary in most instances.

Cements meeting standard specifications for sulfate-resisting Portland cements usually perform well in concrete exposed to alkali sulfate solutions, provided that the sulfate concentration is not excessive and the permeability of the hardened concrete is sufficiently low. However, they do not offer any benefits if under attack by magnesium sulfate solutions, in which the degradation of the C-S-H phase is the dominant cause of degradation.

## 9.2　PRESCRIPTIVE STANDARDS FOR CONCRETE TO BE EXPOSED TO SULFATE ATTACK

As the resistance of concrete to sulfate attack depends not only on the cement employed but also on the mixture proportions, this factor must be taken into consideration in pertinent prescriptive standards. The required parameters of the mixture to be exposed to sulfate may be defined – in addition to the type of cement to be employed – by the water–cement ratio, the amount of cement within the mixture, the strength of the hardened material, or a combination of these. Different requirements may be set depending on the sulfate concentrations in the water or soil in contact with the concrete. The requirements for concrete to be exposed to water containing elevated amounts of sulfates, specified within the standards ACI 201.2R-92 and ACI 318-99 (both identical) and the Uniform Building Code are shown in Tables 1.4 and 1.5. Similar requirements specified by the British Standard specifications, BS 5328, are given in Table 9.1.

*Table 9.1* Recommendations of BS 5328 – Table 7 a–d, Sulfate and acid resistance (BS5328).*

*(a) Recommendations for concrete exposed to sulfate attack*

| Sulfate Class | Exposure conditions — Concentration of sulfate and magensium[1] | | | | | Recommendations | |
|---|---|---|---|---|---|---|---|
| | *In ground water* | | *In soil or fill* | | | *Cement group (from table 7b)* | *Dense fully compacted concrete made with 20 mm nominal maximum size aggregates[2] conforming to BS 882 or BS 1047* |
| | $SO_4$ g/l | $Mg^3$ g/l | *By acid extraction $SO_4\%$* | *By 2:1 water/ soil extract $SO_4$ g/l* | $Mg^3$ g/l | | *Cement content not less than kg/m$^3$* · *Free water –cement ratio not more than* |
| 1 | <0.4 | — | <0.24 | <1.2 | — | 1, 2, 3 | —  — |
| 2 | 0.4–1.4 | — | Classify on the basis of a 2:1 water/soil extract | 1.2–2.3 | — | 1[4]  2  3 | 330  0.50 / 300  0.55 / 280  0.55 |
| 3 | 1.5–3.0 | — | | 2.4–3.7 | — | 2  3 | 340  0.50 / 320  0.50 |
| 4A | 3.1–6.0 | ≤1.0 | | 3.8–6.7 | ≤1.2 | 2  3 | 380  0.45 / 360  0.45 |
| 4B | 3.1–6.0 | >1.0 | | 3.8–6.7 | >1.2 | 3 | 360  0.45 |
| 5A | >6.0 | ≤1.0 | | >6.7 | ≤1.2 | As for class 4A plus surface protection[5] | |
| 5B | >6.0 | >1.0 | | >6.7 | >1.2 | As for class 4B plus surface protection[5] | |

1 Classification on the basis of ground water samples is preferred. Higher values are given for water/soil extract in recognition of the difficulty of obtaining representative samples and of achieving a comparable extraction rate to the indicated by analysis of ground water samples. Suitable methods for the analysis of ground water for sulfate are given in BS 1377: Part 3 and in Building Research Report 279 [6] which also gives methods for determination of magnesium. When results are expressed as $SO_3$ they may be converted to $SO_4$ by multiplying by a factor of 1.2.

*Table 9.1 (continued)*

2 Adjustments to minimum cement contents should be made for aggregates of nominal size other than 20 mm in accordance with Table 8.

3 The limit on water-soluble magnesium does not apply to brackish ground water (chloride content between 12 g/l and 18 g/l).

4 Portland limestone cement should only be used in class 1 sulfate conditions.

5 See CP 102 and BS 8102.

*Note 1* Within the limits specified in this table the sulfate resistance of combinations of ggbs or pfa with SRPC will be at least equivalent to combinations with cement conforming to BS 12, but such combinations are unlikely to exceed the sulfate resigning performance of SRPC.

*Note 2* Cements containing ggbs or pfa are more sensitive to stong magnesium sulfate and a limit on water-soluble magnesium content is given for classes 4 and 5 when using these cements.

*Note 3* The likelihood of attack by sulfate depends on the presence and mobility of groundwater (see Table c and BRE Digest 363 [1]).

---

*(b) Cement groups for use in Table (a)*

| Group | Description |
|---|---|
| 1 | (a) Portland cement conforming to BS 12 |
| | (b) Portland blastfurance cements conforming to BS 146 |
| | (c) High slag blastfurnace cement comforming to BS 4246 |
| | (d) Portland pulverized-fuel ash cements conforming to BS 6588 |
| | (e) Pozzolanic pulverized-fuel ash cement conforming to BS 6610 |
| | (f) Portland limestone cement conforming to BS 7583[1] |
| | (g) Combinations of Portland cement conforming to BS 12 with ggbs conforming to BS 6699 |
| | (h) Combinations of Portland cement conforming to BS 12 with pulverized-fuel ash conforming to BS 3892: Part 1 |
| 2 | (a) Portland pulverized-fuel ash cement conforming to BS 6588, containing not less than 26% of pfa by mass of the nucleus or combinations of Portland cement conforming to BS 12 with pfa conforming to BS 3892: Part 1, where there is not less than 25% pfa and not more than 40% pfa by mass of the combination |
| | (b) High slag blastfurnace cement conforming to BS 4246, containing not less than 74% slag by mass of nacleus or combinations of Portland cement conforming to BS 12 with ggbs conforming to BS 6699 where there is not less than 70% ggbs and not more than 85% ggbs by mass of the combination |

*Note 1* For group 2b cements, granulated blastfurnace slag with alumine content greater than 14% shoulds be used only with Portland cement having a tricalcium aluminate ($C_3A$) content not exceeding 10%.

*Note 2* The nucleus is the total mass of the cement constituents excluding calcium sulfate and any additives such as grinding aids.

| 3 | Sulfate-resisting Portland cement conforming to BS 4027 |

---

1 Poland limestone cement should only be used in class 1 sulfate conditions.

*(c) Modifications to Table (a) for other types of exposure and types of construction*[1]

| | |
|---|---|
| Static ground water[2] | For classes 2, 3 and 4 the requirements for cement group, cement content and free water/cement ratio given in Table (a) may be lowered by one class |
| Basement, embankment or retaining wall | If a hydrostatic head greater than five times the thickness of the concrete is created by the ground water, the classification in Table (a) should be raised by one class. This requirement can be waived if the barrier to prevent moisture transfer through the wall is provided |
| Cast-in-situ concrete over 450 mm thick. Precast ground beams, wall units or piles with smooth surfaces which after normal curing have been exposed to air but protected from rain for several weeks | For classes 2, 3 and 4 and requirements for cement group, cement content and free water/cement ratio given in Table (a) may be lowered by one class |
| | For cast-in-situ reinforced concrete special consideration should be given to the need to maintain adequate cover to the reinforcement |
| Cast-in-situ concrete (other than ground floor slabs[3]) less than 140 mm thick or having many edges and corners | The classification in Table (a) should be raised by one class |

1 Any reductions in sulfate class allowed by this table only apply if other durability and structural considerations permit.
2 Nominally dry sites soils with permeability less than $10^{-5}$ m/s as given in figure 6 of BS 8004: 1986 (e.g. unflassured clay) where it is decided that the groundwater is essentially static (see BRE Digest 363 [1]).
3 For ground floor slabs see BRE Digest 363 [1].

*(d) Modification to Tables (a) and (c) for concrete exposed to attack from acids in natural grounds*

| pH[1] | Mobility of water[2] | Change in classification with respect to minimum cement content[3] and maximum free water/cement ratio for the cement group recommended on the basis of sulfate class in Tables (a) and (c) |
|---|---|---|
| 5.5–3.6 | Static | No change |
| | Mobile | Raise by one sulfate class |
| 3.5–2.5 | Static | Raise by one sulfate class |
| | Mobile | Raise by one sulfate class |

*Table 9.1  (continued)*

1  Department by the method given in clause 9 of BS 1377; Part 3: 1990.
2  See Table (c), note 2.
3  If a cement from group 1 has been selected during the classification for sulfate, when raising by one class in accordance with this table, the cement type may still be used taking as minimum cement content the requirement for group 2 cements.
*Note.* For cast-in-situ or precast culverts see BRE Digest 363 [1].

*Reproduced from BS 5328 Part 1: 1997 with permission of BSI under licence number 2001 SK/ 0090. Complete standards can be obtained from BSI Customer Services, 389 Chiswictk High Road, London, W4 4AL. Tel (+44(0) 20 8994 9001)

## 9.3  PERFORMANCE STANDARDS

In performance tests, specimens of a defined composition and geometry are exposed to sulfate solutions of a defined concentration for a specific time, and the resultant changes are evaluated. The mixture proportions must be defined as they determine the permeability of the hardened material and, thus, the progress of the deterioration process. It is also important to define the geometry and size of the test specimen and, hence, the area of the surface through which the sulfate ions will migrate into the material. Obviously, the mixture has to be pre-cured to a sufficient degree of maturity, which may be defined by a minimum strength of the material, prior to exposure to the sulfate-containing water.

The conditions of exposure of the test specimens to the sulfate solution must be defined by the:

- nature and concentration of the sulfate in the solution to which the test specimens are exposed;
- temperature at which the curing is to be performed;
- number of test specimens and the volume of the sulfate-containing water; and
- duration of exposure to the sulfate solution.

The following modes of exposure to the sulfate solution may be distinguished:

- *Continuous immersion*: The test specimen is continuously immersed in a fixed volume of the sulfate-containing water or in a solution which is periodically renewed, to compensate for the loss of sulfates from solution due to the degradation process. Changes of pH in the solution may be ignored, or the pH may be kept constant by adding hydrochloric acid to the solution as needed. In the course of immersion, the pH of the sulfate-containing solution rises and eventually may attain the pH of a saturated lime solution, 12.2. This does not reflect conditions in the field, where the pH stays constant and at a significantly lower level. This, in turn, affects the rate of degradation that falls as the pH increases.

During the immersion, sulfate ions migrate into the specimen at a rate which depends on the permeability of the material and the concentration of sulfates in the solution. The action of the sulfate solution may result in an expansion of the specimen and/or in cracking and delamination of reacted material from the surface. A well-defined reaction front may separate a layer in which the reaction of the hardened paste with the entered sulfates is virtually complete, while the inner core remains unaffected. Softening of the hardened cement paste, rather than an expansion, may overshadow the usual signs of sulfate attack, if magnesium sulfate acts as the degradative agent.

- *Partial immersion*: The test specimen is only partially immersed in the sulfate-containing water, while the rest is exposed to dry air. Damage tends to occur in areas in contact with air just above the liquid surface, and consists of spalling and scaling. This is brought about both by ettringite-generated expansion and crystallization pressure, as water evaporates from the surface exposed to air.
- *Exposure to wetting–drying cycles*: This form of storage simulates conditions of cyclic migration of sulfate-containing water into concrete. It results in concentration of the sulfate near the concrete surface and leads to an enhancement of the degradative action. The disintegration of the test specimens is a result of both sulfate-induced expansion and crystallization pressure.

Alternatively, one may add excessive amounts of sulfates to the original mixture, rather than storing the test specimen in sulfate-containing water. Under such conditions the changes caused by the sulfates, such as expansion, are uniformly distributed through the material, rather than exhibiting a gradient, with the action of sulfates being greatest close to the surface. This facilitates a quantitative assessment of the action of the sulfates but, at the same time, the permeability of the material is not taken into account.

The extent of damage caused by the excessive sulfates may be assessed:

- on the basis of the appearance of the sample;
- by measuring the length change of the test specimen;
- by determining the strength of the material;
- by determining the changes of mass of the test specimen or by other means.

In selecting the right approach, one has to bear in mind that an expansion of the specimen may be masked by scaling, delamination, and spalling. Also, the strength of the test specimen may first rise (due to filling of the pores with ettringite) before it starts to fall, as cracks develop in the material.

Different steps may be taken to speed up the testing procedure, such as, to increase the concentration of the sulfates in the solution to which the test specimens are exposed, or to increase the temperature, or to apply wetting–drying cycles rather than a continuous immersion in the sulfate-containing water. However, one has to bear in mind that, while the destruction of the test specimens will be accelerated, the mechanism of the destructive action may be altered.

In the ASTM C1012 test, mortar bar specimens, pre-cured to a compressive strength of 20 MPa, are stored in a solution containing 50 g/l $Na_2SO_4$ at 23 °C and the expansion is measured. In a highly sulfate-resistant cement, the expansion must not exceed 0.10% after six months. The method may be used for assessing the sulfate resistance of different types of cement, including blended cements. The pertinent performance specifications for the latter ones are listed in ASTM C1157M. An obvious handicap of the procedure contained in ASTM C1012 is the length of time required to obtain results. Also, the procedure is not directly relevant for sulfate solutions other than alkali sulfates.

In the ASTM C452 test, gypsum is added to Portland cement prior to making the mortar bars, to adjust the $SO_3$ content to 7%. After de-molding, the mortar bars are stored in water at 23 °C and their expansion is measured after fourteen days. The ASTM C150 expansion limit for Type V sulfate resistant cement is 0.040%. The test enables a distinction to be made between Portland cements having different levels of sulfate resistance but it cannot be used for blended cements. The reason is that, with excess of $SO_3$ in the mixture, the action of sulfates starts immediately, rather than after the pozzolanic hydraulic reaction of the added supplementary cementing materials has taken place. Also, this test neither simulates nor predicts field exposure of concrete to sulfates, which involves the ingress of sulfate ions into concrete.

Deficiencies in current standards include lengthy testing periods, the insensitivity of the measurement tools to the progress of sulfate attack, and an uncertain relationship to field degradation mechanism (Clifton *et al.* 1999). Ways to erase these deficiencies are discussed and a proper methodology for developments of new standards is outlined in ASTM E632.

## REFERENCES

ACI 201 (1998) *Guide to Durable Concrete*, ACI Manual of Concrete Practice: Part 1 – 1998, American Concrete Institute, Farmington Hill, MI.

ACI 318-99 (1999) *Building Code Requirements for Structural Concrete*, American Concrete Institute, Farming Hill, MI.

ASTM C150, *Standard Specifications for Portland Cement*, ASTM, Philadelphia.

ASTM C1157M *Standard Performance Specification for Blended Hydraulic Cement*, ASTM, Philadelphia, PA.

ASTM C1012 *Standard Test Method for Length Change of Hydraulic-Cement Mortar Exposed to a Sulfate Solution*, ASTM, Philadelphia.

ASTM E632, *Standard Practice for Developing Accelerated Tests to Aid Prediction of the Service Life of Building Components and Materials*, ASTM, Philadelphia.

BRE Digest 363 (1996) *Sulfate and acid Attack on Concrete in the Ground*, British Building Research Establishment, Garston, Watford, UK.

BS 5328 (1997) "British Standard 5328: Concrete Part 1," *Guide to specifying concrete*, British Standard Institution, Issue 2, May 1999.

Clifton, J.R., Frohnsdorff, G. and Ferraris, C. (1999) "Standard for evaluating the susceptibility of cement-based materials to external sulfate attack", in J. Marchand and J.P. Skalny (eds) *Sulfate Attack Mechanism*, American Ceramic Society, Westerville OH, pp. 337–355.

Hobbs, D.W. (1994) "Minimum requirements for durable concrete, carbonation- and chloride-induced corrosion, freeze-thaw attack and chemical attack", British Cement Association.

Hooton R.D. (1999) "Are sulfate resistance standards adequate?", in J. Marchand and J.P. Skalny (eds) *Materials Science of Concrete Special Volume: Sulfate Attack Mechanisms*. The American Ceramic Society, Westerville, OH, pp. 357–366.

Patzias, T. (1987) "Evaluation of sulfate resistance of hydraulic-cement mortars by the ASTM C1012 Test Method" in *Concrete Durability*, ACI SP-100, American Concrete Institute, Farmington Hills, MI, pp. 92–99.

Patzias, T. (1991) "The development of ASTM C1012 with recommended acceptance limits for sulfate resistance of hydraulic cement", *Cement, Concrete and Aggregates* **13**: 50–57.

Uniform Building Code (1997) "Concrete", Chapter 19, in *Structural Engineering Design Provisions*, vol. 3, International Conference of Building Officials.

# Concluding remarks

The topic of sulfate attack on concrete is a most fascinating one. Because of the chemical and physico-chemical complexities of the mechanisms involved, and the subsequent influence of these processes on the mechanical properties of concrete products and structures, the preparation of this comprehensive review was both exciting and challenging. It was exciting because the intricacies and complexities of the underlying chemistry and physics are intellectually stimulating, requiring complex interpretations, and clear thinking. At the same time, the complexities may lead to honest controversies and stimulating discussions between professional colleagues. While the authors have their own preferences of interpretations, as obvious from the text, they tried to present a multitude of views represented in the literature.

It is our opinion, that neither external nor internal forms of sulfate attack are major causes of degradation of concrete structures. Compared to damage caused by corrosion of steel reinforcement and freezing and thawing, sulfate-related deterioration is globally minute, though serious for the owners of the particular structures. The types of sulfate-related deteriorations that were observed in North America, in the Middle East, and elsewhere are easily preventable by intelligent use of the available knowledge, adherence to the standards and codes, proper materials design, and processing, taking into consideration the expected environment of use.

We hope our readers will find the book interesting and usefull in their professional lives.

# Index